LOCAL ACTION

The New Paradigm in Climate Change Policy

By

Tommy Linstroth

and

Ryan Bell, ICLEI

LIBRARY
FRANKLIN PIERCE UNIVERSITY
RINDGE, NH 03461

University of Vermont Press
Burlington, Vermont
Published by
University Press of New England
Hanover and London

University of Vermont Press
Published by University Press of New England,
One Court Street, Lebanon, NH 03766
www.upne.com
© 2007 by University of Vermont Press
Printed in the United States of America
5 4 3 2 1

All rights reserved. No part of this book may be reproduced in any form
or by any electronic or mechanical means, including storage and retrieval
systems, without permission in writing from the publisher, except by a
reviewer, who may quote brief passages in a review. Members of educational
institutions and organizations wishing to photocopy any of the work for
classroom use, or authors and publishers who would like to obtain permission
for any of the material in the work, should contact Permissions, University
Press of New England, One Court Street, Lebanon, NH 03766.

Library of Congress Cataloging-in-Publication Data

Linstroth, Tommy.
Local action : the new paradigm in climate change policy / Tommy Linstroth and
Ryan Bell. — 1st ed.
      p.   cm.
Includes bibliographical references and index.
ISBN-13: 978-1-58465-672-2 (pbk. : alk. paper)
ISBN-10: 1-58465-672-7 (pbk. : alk. paper)
1. Climatic changes—Government policy.   2. Greenhouse gas mitigation—
Government policy.   I. Bell, Ryan.   II. Title.
QC981.8.C5L56 2007
363.738'74—dc22   2007029121

University Press of New England is a member of the Green
Press Initiative. The paper used in this book meets their
minimum requirement for recycled paper.

LOCAL ACTION

Contents

Acknowledgments

We would like to recognize and extend a special thanks to all who have helped this book come to fruition. These include, but are not limited to, Kristan McKinne and Kurt Schlager for reading the earliest versions and Enid Bell, Gary Bell, Loren Bell, and Debbie Milam for providing input on the final versions. These friends and family provided excellent feedback and much needed support. Without their input this work would have suffered.

We would like especially to thank everyone at ICLEI—Local Governments for Sustainability for all their efforts to work with local governments to reduce greenhouse gas emissions. We also owe them a debt of gratitude for allowing us to make use of data and case studies that they have collected over the years. Their work truly proves the case for effective local action.

Finally, we would like to recognize the efforts of Phyllis Deutsch, the Editor in Chief at the University Press of New England, for guiding us through the publication process. Similarly, we truly appreciate the efforts of Melissa Capria and Dr. Eban Goodstein for their final reviews and excellent input to help create a better final product.

We dedicate this book to our immediate families: our parents Mike and Jean Linstroth and Marilyn and Gary Bell, and to Brittni and Natasha Milam Bell. Without their ongoing support for us and our aspirations over the years, and their help in the last weeks of this project, we never would have reached a point of putting a pen to paper on this topic.

Introduction

We are all familiar with the horror stories about where we started from thirty years ago. As we entered the 1970s, the Nation's waters were in crisis—the Potomac River was too polluted for swimming, Lake Erie was dying, and the Cuyahoga River had burst into flames. Many of the Nation's waterways were little more than open sewers.

The 1972 Clean Water Act has sharply increased the number of waterways that are once again safe for fishing and swimming. The Act launched an all-out assault on water pollution, and it worked well. It enabled us to improve water quality all across the nation while experiencing record economic growth and a sizeable expansion of our population.[1]

> G. Tracy Mehan III,
> Assistant Administrator for Water, U.S.
> Environmental Protection Agency, statement
> before the Committee on Environment
> and Public Works, October 8, 2002

A major problem with public policy, and especially environmental policy, is that it is reactionary. Too often things already have gone awry, or are well on their way, before legislators enact new laws to protect the populace and to prevent things from getting worse. The Clean Water Act was not passed because policymakers foresaw future degradation of our natural water resources. The destruction was already at hand, with the majority of our nation's waters unsuitable for swimming and fishing. The same holds true for the Clean Air Act, the Comprehensive Environmental Response, Compensation, and Liability Act (the Superfund law), the Endangered Species Act, and most major environmental legislation of the twentieth century. Fortunately, these powerful pieces of legislation were passed in time, and air quality, the health of our rivers, lakes, and streams, and the populations of many species have rebounded significantly in the United States. The water is cleaner, the air is purer, and species are being preserved (for the time being). However, the nation had to experience how dire the situation could become before having the prudence to pass legislation to make it better.

Global warming is another such issue facing the nation today, although this generation is not solely responsible for greenhouse gas emissions. Humans have been contributing to global warming since the dawn of the industrial revolution, when the widespread burning of fossil fuel as a cheap energy source began to pour carbon dioxide and other pollutants into our atmosphere. These gases have been accumulating and surface temperatures have been increasing for the last two hundred years, and most scientists believe that we are at a critical junction. Strong action must be taken to reverse historic trends for the sake of current and future generations. The impacts already are being felt around the country and the world, and within decades it will be too late to reverse the warming trends.

Unfortunately, the call for the United States to take action on climate change to date, has not been heeded by policymakers. The United States, which has the economic strength and technology to lead the world in so many matters, has remained staunchly apathetic about the potential ramifications of global warming.

There are many reasons for this inaction. Global warming is seen as a partisan issue, effectively removing half of the population from the debate. Unrealistic calls to stop using fossil fuels immediately and switch to renewable power sources polarize discussions. Perceived harm to the economy—by far the number one reason that administrations have hesitated to initiate change or ignored the issue altogether—is still a prevalent concern to policymakers. Moreover, until recently, public understanding of, and support for, global warming action has been nonexistent.

This book differs from other climate change policy books in two significant ways. First, it does not call on the federal government to pass sweeping legislation such as was done for the Clean Air Act and Clean Water Act. While this step certainly would be welcome, this book's focus is on another, more flexible solution that already is being implemented across the country: local government action to reduce climate change emissions at the city and county level.

This grassroots approach to emissions reduction is a powerful tool for combating climate change. The majority of U.S. citizens live in cities, and therefore a vast majority of domestic emissions are generated within cities or in order to service their populations. Transportation and traffic are heavier

in cities, while urban residents and the commercial and industrial enter-
prises on which they depend consume vast amounts of electricity and fuels.
The massive landfills that service most urban areas are a significant source of
greenhouse gases more powerful than carbon dioxide. Even the physical
materials that go into constructing urban areas (cement, asphalt, and wood)
have an impact on the delicate balance of gases in the atmosphere. We need
to reduce emissions where they start—in cities—and it is up to local governments to take action.

While a federal policy may end up being far reaching and effective, it is
not a quickly implemented solution. More global warming legislation has entered Congress in recent years than ever before—yet none has been transformed into a binding law. Working at the local level can eliminate much
of the political posturing that takes place in the federal government, and
policies can be implemented much more rapidly. While federal legislation
also targets specific industries, local action allows individual governments to
identify specific causes for concern in their communities, be they transportation, waste management, building codes, or energy use. These specific
target areas may be different for each of the thousands of local governments
across this nation, allowing for flexible implementation customized for specific needs.

Second, this book is not based on abstract theories. It uses real-life case
studies to demonstrate different strategies that local governments are pursu-
ing to reduce their local greenhouse gas emissions. The reward of being a
first mover comes with considerable risk. As such, the strategies highlighted
in this book have been chosen for their proven results. The solutions often
can be implemented easily, and they not only reduce greenhouse gas emissions, but also save taxpayer money and increase the quality of life for residents. The local governments that have adopted these measures afford a
proven resource for others as they strive to implement effective climate
change policies at the local level.

This book is broken up into nine chapters. Chapter One provides a brief
overview of global warming. It identifies current trends and the gases that
contribute to climate change, and discusses why it is indeed a salient issue in
the twenty-first century. Chapter Two explains the United States's history of
action (or inaction) on the issue of climate change and describes the federal
policy options currently under discussion. Chapter Three builds the case for
local action by detailing how local governments can have profound impacts
on global issues such as climate change and listing the additional benefits

that a jurisdiction reaps from taking action to reduce emissions. These benefits include improved air quality, decreased traffic congestion, increased citizen health, reduced operating costs, and an overall improvement in quality of life. Chapter Four identifies the constraints faced by local policymakers who want to address global warming. Chapters Five and Six begin to examine some of the more popular and effective measures that municipalities have implemented to decrease local greenhouse gas emissions. Chapter Five focuses on community-based initiatives, while Chapter Six concentrates on measures within local governments' operations. Chapters Seven and Eight are case studies of two of the U.S. communities that are leaders in reducing greenhouse gas emissions at the local level. Finally, Chapter Nine discusses where we can go from here—both through local action and through federal climate change policy.

There is the need for one disclaimer. Although this text presents the most accurate and up-to-date information possible, this is a rapidly evolving field and therefore a host of issues may have changed by the time this book is in print. Most immediately, it is hoped that bills will continue to be introduced into Congress addressing climate change, more local governments will take action, and less greenhouse gases will be emitted. In this crucial area, action is needed desperately. The authors applaud any positive developments, even if they do compromise the inclusivity of this book.

LOCAL ACTION

Chapter One

A Global Warming Overview

Unless we abate the growing threat of global warming pollution, everything we do today to make a better world for tomorrow will be for naught.[1]

Mayor Robert E. Minsky, Port St. Lucie, Florida

For the past twenty-five years, serious scientific discussions have been conducted on global warming and on the role that humans play in altering Earth's climate. Recently, U.S. citizens have seen years of drought, years of flooding, changing weather patterns, heat waves, increasing frequency of El Niño events, and a host of other climatic anomalies. Although individually these cannot be tied directly to a changing climate, taken together they point toward a disturbing trend. Glaciers are receding around the world. The frequency of storms during the 2004 hurricane season and the disastrous storms that battered the Gulf Coast in the summer of 2005 are consistent with what is expected as ocean temperatures increase due to a warming climate. Heat waves during the summer of 2006 killed more than a hundred people in California alone, and each consecutive year seems to be one of the warmest years on record. In fact, the fifteen warmest years since record-keeping began in the mid-1800s have occurred since 1980.[2] Record-breaking global temperatures continue to occur regularly. Preliminary data show that 2006 will have been the warmest year on record, replacing 2005, 1998, 1997, and 1990 respectively.[3] Indeed, new studies released in February of 2007 show that eleven of the twelve years between 1995 and 2006 will be the warmest years recorded.[4]

Change has been a constant throughout Earth's history. Sea levels rise and fall, glaciers advance and retreat, rain falls, the sun scorches, volcanoes erupt, the Earth's crust breaks and shifts. These natural processes have been happening over thousands of years. The Earth's climate also has displayed variability, with climate change happening on various scales throughout history. So what exactly is meant by the term "global climate change" in current debates and discussions?

While the definition holds many meanings in the political realm, in the scientific community, climate change refers to "any change in climate over time, whether due to natural variability or as a result of human activity."[5] Variations in climate are influenced by a variety of natural factors including changes in the Earth's orbit, volcanic activity, solar irradiance, and the composition of the atmosphere.

The Greenhouse Effect

The role of atmospheric gases in regulating global temperatures has been well established since originally suggested by Joseph Fourier in 1924. Under natural conditions, the Earth is bombarded continually by energy from the Sun. Most of this strikes the Earth in the form of short-wave solar energy (e.g., ultraviolet and visible light). When this energy strikes the surface of the Earth, it is absorbed, converted to long-wave radiation (e.g., infrared energy or heat), and radiated back into space. Some gases in the atmosphere allow the short-wave radiation from the Sun to pass through but trap the escaping heat (long-wave radiation) close to the surface of the Earth. These are called greenhouse gases because they blanket the Earth, functioning in a similar fashion to a farmer's greenhouse whose glass walls are transparent to light but opaque to heat energy. This results in temperatures within the greenhouse being significantly warmer than outside temperatures. These greenhouse gases form a blanket around the planet which keeps the Earth's surface about 54° F warmer than it would be without the influence of these gases. Currently, average global temperature is approximately 55 to 60°F due to the naturally occurring greenhouse effect. Without these greenhouse gases, this temperature would be closer to 0°F.[6]

This phenomenon, known as the "greenhouse effect," is necessary to support life on Earth. The current ecologic systems that exist around the planet are dependent on temperatures being maintained within a fairly narrow range. If this blanket of gases is too thin, the atmosphere of the Earth will more closely resemble that of Mars, which traps very little heat, keeping the surface temperature well below freezing. On the other hand, too high a concentration of greenhouse gases brings the composition of the atmosphere closer to that of Venus, which is too hot to support life.[7]

The atmosphere surrounding the Earth is a complex mixture of gases held to a relatively narrow range of concentrations (see Table 1). The principal

Table 1. **Concentrations of Gases in Earth's Atmosphere**

Gas	Concentration in the Atmosphere (%)
Nitrogen (N_2)	78.08
Oxygen (O_2)	20.95
Water* (H_2O)	0 to 4
Argon (Ar)	0.93
Carbon dioxide* (CO_2)	0.0360
Neon (Ne)	0.0018
Helium (He)	0.0005
Methane*(CH_4)	0.00017
Hydrogen (H_2)	0.00005
Nitrous oxide (N_2O)	0.00003
Ozone (O_3)*	0.000004

*Atmospheric concentrations can vary

SOURCE: M. Pidwirny, *Fundamentals of Physical Geography*, 2nd edition, 2006. Chapter 7a, www.physicalgeography.net/fundamentals/contents .html (accessed on December 13, 2006).

naturally occurring greenhouse gases are carbon dioxide (CO_2), methane (CH_4) and water vapor. Although these are only present in minute concentrations, they have a significant impact on regulating global temperatures.

CO_2 is the primary greenhouse gas of interest, as human activities are greatly increasing its concentration.[8] Therefore, it is important to understand the naturally occurring carbon cycle when discussing the greenhouse effect. Normal background concentrations of CO_2 are somewhat variable from year to year but generally remain constant over the course of millennia. The concentration of CO_2 is regulated by the various biologic, geologic, and atmospheric forces that come into play in the carbon cycle.

The most widely understood part of the global carbon cycle has to do with the movement of carbon through biological systems. In this cycle, CO_2 is sequestered from the atmosphere by plants through the process of photosynthesis. Vegetation removes CO_2 from the atmosphere, incorporates the carbon into its physical structures, and releases the oxygen back to the atmosphere. In turn, carbon dioxide is released when that vegetative material is broken down. This could be through natural decomposition, fire, and/or respiration.[9] This part of the carbon cycle occurs in a fairly rapid fashion and has a relatively small impact on the global concentration of CO_2 in the atmosphere. Over short time scales, approximately the same amount of CO_2 is removed from the atmosphere through photosynthesis as is returned by other means (decomposition, etc.).

Other natural carbon cycles occur on much longer "geologic" timeframes, measured in millennia as opposed to decades. In some cases, plants and other organic materials do not decompose as described above; rather they are deposited in conditions that inhibit decay (such as in peat bogs). This vegetative material can then become buried and, through pressure and heat applied over millions of years, be converted into coal, oil, natural gas, and other fossil fuels—so called because the carbon embodied in them has become fossilized in the geologic layers of the Earth.

Another geologic carbon cycle involves carbon dioxide dissolving or being released from the oceans depending on temperature or atmospheric concentrations. Once in the oceans, the carbon can be used by marine plants or by animals to form shells and coral reefs. When these organisms die, their shells sink to the sea floor, forming layers of sediment that are converted over time into solid rock. The carbon embedded in these carbonaceous rock layers is stabilized and is generally removed from the carbon cycle unless released in relatively minor amounts by geologic forces such as weathering or volcanic eruptions.

All aspects of the global carbon cycle work together to keep the atmospheric concentration of greenhouse gases in a relatively steady state at or near the concentrations shown in Table 1. However, human activities can affect these concentrations, and this is of primary concern to scientists and policymakers. Human intervention in these systems (such as the extraction and combustion of fossil fuels) accelerates the release of stored carbon into the atmosphere to a rate faster than it can be removed by natural systems. Human activity also causes carbon to be released in a form other than CO_2 (detailed later) that can trap a much greater amount of heat close to the

Earth's surface. Such activities result in rising temperatures or "enhanced greenhouse effect," leading to climate change or global warming.

Climate Change and Greenhouse Gases

In the past century, the concentration of carbon dioxide in the atmosphere has risen almost 30%, methane levels have more than doubled, and nitrous oxide concentrations have increased by 15%.[10] Furthermore, the growth in atmospheric greenhouse gas concentrations has been progressing at a more rapid rate over time. For example, concentrations of CO_2 have increased by 54 parts per million (ppm) over the last forty years (1960–2000), a rate that far surpasses the 36 ppm increase that occurred during the previous two hundred years (1760–1960).[11] Models indicate that the majority of this increase is the direct result of the emissions released during the burning of fossil fuels and other human activities.

Increasing the concentration of greenhouse gases in the atmosphere is equivalent to creating a thicker blanket around the Earth, which can trap more heat and therefore increase temperatures at the planet's surface. Atmospheric scientists widely agree that this is exactly what is happening and that it is specifically due to human activity that has occurred since the industrial revolution transformed society and its requirements for fossil fuel–based energy sources in the early 1800s.[12] Overwhelming evidence suggests that the accumulation of greenhouse gases in the atmosphere is responsible for the Earth's rising temperature.[13] The major source of anthropogenic greenhouse gas emissions is the burning of fossil fuels by industry and for transportation, electricity, heating, and cooling. The combustion process releases the carbon contained within the fuel as CO_2. This carbon was removed from the atmosphere by photosynthesis millions of years ago. Therefore, this CO_2 production is above and beyond what normally would be released through the carbon cycle, and there is no natural path for removing it. As a result, the use of fossil fuels increases the overall concentration of greenhouse gas in the atmosphere.

The idea that the release of greenhouse gas from human activity could lead to an increase in global temperatures is not a new one. By the mid-1800s, scientists realized that various gases, common in the atmosphere, were transparent to light but prevented the transfer of heat. As early as 1906, the Swedish scientist Svante Arrhenius, a Nobel Prize–winning chemist, proposed his "hot house theory" that postulated that human sources of greenhouse gases could cause global temperatures to increase by 9°F. This prediction is remarkably close to current projections from the Intergovernmental Panel on Climate Change (IPCC), the world's largest international collective of climate experts.[14] After analyzing global climate models, the IPCC reported in 2001 that global temperatures are likely to increase by 1.98 to 11.52°F by 2100.[15]

Many gases released by human activity contribute both directly and indirectly to the greenhouse effect. The most important ones, which have been identified and focused upon by the international scientific and political communities as the emissions that should be reduced to curb the "enhanced greenhouse effect" include:

- Carbon dioxide (CO_2)

- Hydrofluorocarbons (HFCs)

- Methane (CH_4)

- Nitrous oxide (N_2O)

- Perfluorocarbons (PFCs)

- Sulfur hexafluoride (SF_6)

The U.S. Environmental Protection Agency (EPA) and the Intergovernmental Panel on Climate Change describe these greenhouse gases and their changes in atmospheric concentrations as follows.

Carbon Dioxide (CO_2). As described previously, carbon dioxide is the most prevalent of the greenhouse gases. CO_2 can move into and out of the atmosphere from carbon sources in the ocean, flora and fauna, and from geologic reserves. Human activity has caused atmospheric CO_2 concentration to increase by 31% since 1750 (pre-industrial revolution). Current levels exceed 379 ppm (in 2005), a situation that likely has not occurred naturally in the past 650,000 years (and perhaps the past twenty million years). Between 70 and 90% of anthropogenic CO_2 emissions result from the burning of fossil fuels, with the remainder being released from land-use changes, primarily deforestation, which has the double impact of a loss of the forests' sequestration potential and the release of carbon embedded in the trees themselves.

Methane (CH$_4$). The predominant source of methane in the atmosphere is from the decomposition of organic material. When objects decay in conditions where oxygen is severely limited, such as in landfills and sewage treatment plants, the carbon released combines with hydrogen to form methane as opposed to carbon dioxide. Methane is also released from incomplete fossil fuel combustion, leaking natural gas infrastructure, and a byproduct of agricultural practices, such as cattle and rice production. The atmospheric concentration of CH$_4$ has increased 148% since 1750 and continues to increase. The present atmospheric concentration has not been exceeded during the past 650,000 years.

Nitrous Oxide (N$_2$O). Nitrous oxide is also produced by the combustion of organic materials and fossil fuels (especially in automobiles), as well as from agriculture processes (especially the use of fertilizers), wastewater treatment, and some industrial processes. N$_2$O concentrations have increased 18% since 1750, and current concentrations of N$_2$O in the atmosphere are higher than any time in the last thousand years.

Hydrofluorocarbons (HFC), Perfluorocarbons (PFC), and Sulfur Hexafluoride (SF$_6$). These complex compounds are produced for industrial and manufacturing processes, refrigerants, and as replacements for ozone-depleting substances. Aluminum smelting, electric power distribution, and magnesium casting are large sources of these compounds. These chemicals are powerful greenhouse gases and remain in the atmosphere for long periods of time. As these gases are largely not naturally occurring, their use by humans leads to increased atmospheric concentrations.[16]

Other greenhouse gases are released by human activity. These include ground-level ozone, carbon monoxide (CO), nitrogen oxides (NO$_x$), and various volatile organic compounds. Although concentrations of these gases have increased in some areas, they have relatively short life expectancies in the atmosphere and are easily washed out by precipitation events or simply break down rapidly. Therefore they are not considered as critical to a discussion of long-term global climate change. Similarly, some halocarbons are powerful manufactured greenhouse gases, but they have been outlawed (due to their ozone-depleting properties) and are generally declining in atmospheric concentration. For this reason, they also are not considered to be of critical importance (although the same cannot be said for their replace-ments, which were discussed above).[17]

These greenhouse gases capture heat energy to varying degrees as ex-pressed by the gases' global-warming potential, which is a measurement of the impact of a gas on global warming. The global-warming potential of a gas is dependent on how it reacts with long-wave (infrared/heat) radiation coming from the Earth and how long it remains in the atmosphere. Gases that trap a large amount of heat but remain in the atmosphere for a relatively short period of time may have a lower global-warming potential than gases that trap less heat but remain in the atmosphere for a longer period of time.

The concept of global-warming potential allows different greenhouse gases to be converted to a common unit based on their heat trapping capacity. In the United States, the units used to express amounts of different greenhouse gases are "tons of carbon dioxide equivalents" (tons CO_2E or tons eCO_2).[18] A carbon dioxide equivalent is the amount of CO_2 it would take to trap the same amount of heat as a given quantity of the greenhouse gas in question. This value is the product of the weight of the gas in tons and its global-warming potential.

For example, one molecule of SF_6 warms the planet to a similar extent as 23,900 molecules of CO_2; therefore the global-warming potential of SF_6 is 23,900. In this example, 1 ton of SF_6 could also be expressed as 23,900 tons of eCO_2. Common greenhouse gases and their respective global-warming potentials are listed in Table 2.

The benefit of using global-warming potentials to convert greenhouse gases into eCO_2 is that it allows for a quick comparison of different gases rel-ative to the effect they have in the atmosphere. Without this standard for comparison, it would be difficult to compare the impact that diverse activi-ties have on climate change. For example, which has more of an impact on global warming—the carbon dioxide released from the operation of a mu-nicipal fleet or the methane released from the municipal landfill?

Effects of a Changing Climate in the United States

While it is impossible to predict exactly what the impact of a rapidly warming planet will be, it is clear that there will be significant changes that have implications throughout the country. Although the impacts of climate change will vary by region and in their intensity, very few (if any) regions will escape being affected. The predicted changes are not limited to an increase

Table 2. **Atmospheric Lifetimes and Global-Warming Potentials of Common Greenhouse Gases**

Gas	Atmospheric Lifetime (Years)	Global-Warming Potential (100 Year)
Carbon dioxide (CO_2)	50–200	1
Methane (CH_4)	9–15	21
Nitrous oxide (N_2O)	120	310
Hydrofluoro-carbons (HFC)		
HFC-23	260	11,700
HFC-125	29	2,800
HFC-134a	13.8	1,300
HFC-143a	52	3,800
HFC-152a	1.4	140
HFC-227ea	33	2,900
HFC-236fa	22	6,300
Perfluorocarbons		
Perfluoromethane (CF_4)	50,000	6,500
Perfluoroethane (C_2F_6)	10,000	9,200
Sulfur hexafluoride (SF_6)	3,200	23,900

SOURCE: Intergovernmental Panel on Climate Change, *Climate Change 1995: The Science of Climate Change*. (Cambridge, U.K.: Cambridge University Press, 1996).

in temperature; rather the consequences will be widespread as changing temperatures will also have an affect on issues as diverse as local hydrologic regimes, diseases and human health, economies, local ecosystem structure, and many other aspects of the world we know. Although there is the potential for some regions to undergo what some might consider positive changes (for example, longer growing seasons) adjusting to the new conditions that climate change will bring represents a shift from the status quo and will require significant modifications in existing social, economic, and infrastructure systems that will necessitate the investment of time, money, and other resources.

The remainder of this chapter is devoted to highlighting some of the potential impacts that climate change will have on selected regions of the United States. Although this is not a comprehensive look at the entire country, the same basic predictions are repeated time and again, regardless of what region is being considered.

According to the EPA, the forty-four million people living in the Southeastern United States are "particularly vulnerable" to the impacts that global climate change will bring to the region.[19] Everyone in this fast-growing region of the country would be affected by the impacts of hurricane seasons with increased storm intensity and frequency that are predicted by the IPCC. For a region pummeled by numerous hurricanes in 2004 that resulted in billions of dollars of damages, this is a threat that should not be ignored. If sea levels were to rise by only 20 inches, 35 to 50% of the region's coastal wetlands would be destroyed and millions of people would be forced to relocate. Losses from the encroaching seas would include many ecologically important habitats such as the Everglades and Florida Keys. Additionally, increasing temperatures allow tropical diseases such as malaria to move into the region, and increase harmful smog levels in cities such as Atlanta that are already fighting to improve their air quality. These outcomes of global warming will translate into an economic hardship for the region. Already, economic losses from natural disasters were eight times greater during the 1990s than the 1960s.[20]

The arid southwestern states contain some of the fastest-growing urban areas in the country. The geographic diversity of the region makes it difficult to generalize on the outcomes of global warming, although a historic trend of a rising temperatures holds true throughout the region.[21] For example, average temperatures in Tucson, Arizona, have increased by 3.6°F in the last one hundred years, and could rise by another 3 to 5 degrees in the next century.[22] These warmer temperatures will have a large impact on energy demand in a region that already has high cooling costs and will add to the heat stresses that already affect many residents living in urban areas.[23]

Fluctuations in rainfall events will have the potential of increasing droughts and instances of flooding, landslides, and erosion at different times and locations.[24] According to the EPA, California has already seen a 20% decrease in precipitation in many parts of the state.[25] Although the long-term impacts of climate change on precipitation are likely to vary by region, a greater portion likely will fall as rain as opposed to snow, leading to negative impacts on winter tourism, available water resources in the summer, and hydropower generation potential.[26]

In addition, diseases, especially Hantavirus and those diseases transmitted by mosquitoes, are likely to increase in prevalence. Wildfires are likely to increase in occurrence and intensity. Coastal infrastructure also will be increasingly threatened by sea-level rise and the associated coastal inundation.[27]

Predictions by the EPA and the Climate Impacts Group (which focuses on climate change and the Pacific Northwest) hold that the northwestern United States should expect year-round temperature rises of between 1.6 and 6.3°F. As the climate warms, the snow pack will shrink and summer stream flows will drop considerably. This would lead to increased winter flooding and low summer river levels. This shift in the hydrologic regime will adversely affect important fish habitat, including that of already-endangered salmon populations. The region also will be affected by rising sea levels, erosion, and saltwater encroachment into drinking water supplies.

Human health risks also increase with a rise in temperature. In Portland, it is estimated that a 4 degree summer temperature increase will cause heat-related mortalities to jump 150% and in Seattle a 3 degree winter temperature increase would cause a 160% increase in mortality due to more inclement weather. In addition, the iconic giant forests of the Pacific Northwest (which are also an economic driver in the region) will face drought stress and increased susceptibility to pests and wildfires, generally decreasing in range by 15 to 25%.[28]

The Mid-Atlantic region is home to the nation's capitol, is responsible for 13% of the nation's economic output, and contains the nation's largest estuaries—including the Chesapeake Bay. Approximately 50% of the region's population lives along its extensive coastlines. This makes the Mid-Atlantic particularly vulnerable to sea-level rise, hurricanes and other storms, seawater encroachment into drinking water supplies, and costal erosion. The area's ecosystems also are likely to face drastic modification from changes in temperature and rainfall. The species composition of the region's deciduous forests (which cover 65% of the region) is liable to become less diverse and shift from maple, beech, and birch to southern types of oak, hickory, and pine. This also will have an effect on the migratory birds that pass through the region and rely on its habitats for food and shelter. Similarly, as water temperatures increase, trout and other cold-water species will be replaced by warm-water species. All ecosystems will become more susceptible to pests and invasive species.[29]

Once again, warmer climate patterns will exacerbate human health problems. The Mid-Atlantic region already has an elevated rate of heat-related mortality; higher temperatures could lead to 400 to 500 more deaths per year in Philadelphia and other large cities. Similarly, increased rainfall and floods would cause more outbreaks of illnesses that are spread by contaminated water. Although these health risks will affect the entire population, at-risk groups such as children, the elderly, those in poor health, and individuals without access to air conditioning, clean water, or health care (i.e., the economically disadvantaged) are likely to be disproportionately impacted.[30]

The world's largest reservoir of fresh water is the Great Lakes. This region of the United States and its delicate ecosystems face the pressures of climate change as well. A predicted warming of 4 to 8°F and a 25% increase in annual precipitation will cause shifts in the ranges of terrestrial species. Although rain events will tend to become more intense, temporarily raising local water levels, the long-term trend will be a decline in lake levels of 1.5 to 8 feet due to decreased snowfall and the increased evaporation rates that accompany warmer weather. These changes have serious implications for regional ecosystems and economies based on coastal habitats that are unable to adapt to a rapidly changing environment. Populations of migratory neotropic song birds, as one indicator, are expected to drop by 20 to 67% (depending on the state under consideration).[31]

The East Coast region of the United States contains the New York City metropolitan area. These thirty-one counties contain approximately 600 miles of waterfront, four out of the five boroughs are located on islands, and nearly 2,200 bridges and tunnels link the region together. As such, the region is particularly susceptible to sea-level rises. Much of the transportation system (tunnel and subway entrances, and all three airports) are located less than 10 feet above sea level. Besides consequences of flooding, climate change could bring water shortages, dangerously hot summers (as seen in 2006), and increased mosquito-borne diseases and air pollution problems.[32]

Other regions of the country are expected to face similar consequences as well, including declining snow packs, changes in rain patterns, species loss, and health concerns. If current predictions are realized, everyone, no matter where they live, will have deal with the impacts of climate change. With the risks of global warming becoming evident around us, one would believe that the potential for widespread catastrophic disaster would prompt the federal government to take aggressive action to reduce greenhouse gas emissions. This is especially true as the United States, with 3% of the world's population, is responsible for 25% of planet's total anthropogenic CO_2 emissions.[33] However, as chapter Two iterates, this has not been the case to date.

Chapter Two

History of U.S. Climate Change Policy

If the U.S. government continues to ignore the rest of the world on energy, then it is time for all local leaders to speak out.[1]

Mayor John C. Rayson, Pompano Beach, Florida

Environmental problems in the early twenty-first century have begun to occupy a more prominent position on national and international political agendas, although debate still continues over their severity, the level of governmental intervention needed, and the policy approaches that work best. As mentioned in the introduction, environmental policy in the United States tends to be reactionary, addressing problems only after they have already become very apparent. Unfortunately, the evident existence of environmental degradation does not ensure that the problems will reach the public policy agenda, much less do abstract predictions of future environmental degradation that has yet to manifest itself in irrefutable impacts (as is the case with climate change).[2] As yet, the U.S. government has not established a national policy to decrease its domestic greenhouse gas emissions, instead choosing to research the issue further and allowing emissions to increase with economic growth by establishing a greenhouse gas reduction target tied to gross domestic product. This lack of action on climate change has been exacerbated further by the national focus on terrorism and homeland security that arose following September 11, 2001, pushing other concerns off of policymakers' agendas.[3]

The United States' official stance on global warming has been conservative since the issue first appeared in the international policy arena. "The American way of life is not negotiable," stated President George H. W. Bush at the 1992 World Environmental Summit in Rio de Janeiro.[4] That statement, made in response to the call for Western nations to change their lifestyles to prevent further global warming, established the course for U.S. federal climate change policy. It was at this summit that the United Nations Framework Convention on Climate Change (UNFCCC) was adopted, calling on industrialized nations to voluntarily reduce emissions to 1990 levels by 2000. However, President Bush's remarks foreshadowed fifteen years of U.S. climate change policy that emphasized inaction over regulations or any other alteration to "business as usual."

Although the theory of anthropogenic global warming has been around since the beginning of the twentieth century, it fully entered the world stage as an issue of international concern during the 1980s. In 1988, the World Meteorological Organization and U.N. Environmental Programme established the International Panel on Climate Change (IPCC) to assess scientific, technical, and socioeconomic information related to human-caused climate change.[5]

By 1990, the IPCC was able to report that concentrations of anthropogenic greenhouse gas emissions have been increasing since the industrial revolution and that this would lead to an increase in global temperatures, a position that has been strengthened with each subsequent report.[6] This finding began the dialogue around the need for the world's governments to take immediate actions to reduce each country's greenhouse gas emissions. The United Nations General Assembly called for the development of an international convention (which would become the UNFCCC) on climate change prior to the 1992 Environmental Summit in Rio.[7]

In 1988, the U.S. Senate also began to hold hearings on climate change, at which NASA's chief scientist James Hansen announced his position that the warmer temperatures being recorded were a symptom of anthropogenic climate change.[8] Vice President (and presidential candidate) George H. W. Bush also took a strong stand later that year, stating that "those who think we are powerless to do anything about the greenhouse effect are forgetting about the power of the White House effect. As President, I intend to do something about it."[9] However, those strong statements were the extent of the administration's commitment to reduce greenhouse gas emissions. As demonstrated in his aforementioned comments during the World Environmental Summit, President Bush was not willing to take any concrete steps to reduce U.S. greenhouse gas emissions, especially if there was a perception that they would hurt the economy. Indeed, the policy taken by the United States in the following years was to resist calls for any internationally binding commitments or timetables for the reduction of greenhouse gas emissions.[10]

Despite this opposition, the 1992 World Environmental Summit in Rio de Janeiro generated a serious international dialogue on climate change, which culminated with the adoption of the UNFCCC. Although it did not call for specific national limits on greenhouse gas emissions or contain enforcement provisions (largely due to the U.S. delegation's opposition), the UNFCCC set up a mechanism for governments to work together to tackle the issue of climate change. Its stated objective is the "stabilization of greenhouse gas concentrations in the atmosphere at a level that would prevent dangerous anthropogenic interference with the climate system," within a time frame that "allow[s] ecosystems to adapt naturally to climate change, to ensure that food production is not threatened, and to enable economic development to proceed in a sustainable manner."[11]

Governments signing the framework convention agreed to meet together annually and to share information on emissions, policies, and best practices; put strategies for reducing emissions into place; and work together to adapt to the impacts of climate change.[12]

While at the Earth Summit, the United States signed the UNFCCC, and the Senate ratified it on October 7, 1992.[13] The convention established the ultimate objective of stabilizing greenhouse gas concentrations at 1990 levels. By March of 1994, enough countries had ratified the treaty for it to enter into force.[14] Despite almost universal acceptance of the UNFCCC and its commitments to take action to curb emissions, it was not until 1997, in Kyoto, Japan, that the signatories to the treaty were able to work out their differences and draft an agreement that established specific targets and mechanisms (in the form of the Kyoto Protocol). Even then, it was not until 2005 that the protocol received the signatures required to enter into legal force (without the United States' participation)—thirteen years after the Earth Summit and seventeen years after the formation of the IPCC.

The U.S. Position through the 1990s: The Clinton Administration

When William Jefferson Clinton defeated incumbent George H. W. Bush, there was initial hope that the United States would embark on a new policy course to address global warming; however, few tangible steps were taken. The Clinton administration's lack of climate policy does not necessarily reflect a lack of effort. During February 1993, President Clinton's first full month in office, his administration began to look at the possibility of instituting a carbon tax based on the heat content of all fossil fuels (as measured in British Thermal Units). This tax was expected to reduce carbon emissions through increased energy conservation and more efficient energy consumption and to generate revenue to pay down the national deficit.[15] In his Earth Day speech on April 21, 1993, President Clinton announced "our nation's commitment to reducing our emissions of greenhouse gases to their 1990 levels by the year 2000."[16] The tax, which would have resulted in a tax of 7.5 cents on gasoline and $5.35 per ton of coal, passed the House but was ultimately defeated in the Senate, after aggressive lobbying by the energy, farming, and aluminum industries.[17] Later that year, a much less stringent "Transportation Fuels Tax" was passed that levied a 13.814-cent surcharge on gasoline and diesel fuel, but exempted heating fuels, exported fuels, international aviation and shipping fuels, and fuels used off-road, on farms, and by state and local governments or nonprofit organizations.

In response to the failure of the carbon tax, the administration pursued an alternate strategy that focused on providing incentives and supporting research. This strategy, which became known as the Clinton-Gore Climate Change Action Plan, was soon adopted.[18] This plan included approximately fifty voluntary measures to increase energy efficiency, utilize renewable energy resources, and facilitate tree planting to absorb carbon dioxide.[19] While initially popular, enlisting more than five thousand companies and organizations, the program has not had much success in actually reducing national greenhouse gas emissions. In fact, low fuel prices and the rapid economic growth experienced during the 1990s caused emissions to increase by 13% during the term of President Clinton—a total of 15% over 1990 levels.[20]

On a global level, there were several follow-up sessions to the UNFCCC. These annual meetings, known as the Conferences of the Parties (COP), bring together signatories to the UNFCCC to discuss implementation and next steps. The first COP was held in Berlin, Germany, in March 1995, where the participating nations issued the Berlin Mandate. The mandate acknowledged that the voluntary approach to emissions reductions had failed, and that for meaningful reductions to occur, industrialized countries would have to make binding commitments to reduce their greenhouse gas emissions.[21]

During July 1996, nations met in Geneva, Switzerland, for COP-2. During the meetings, the United States announced support for binding commitments and timetables for greenhouse gas reductions (although not before 2010).[22] This was seen as a positive step for the country, which had been opposing binding commitments since the international dialogue on climate change began in 1992. However, the United States then argued for multi-year targets, pacing of emissions reductions, implementation of an emissions trading program between developed nations, and for the ability to take credit for programs that industrialized countries launch in undeveloped nations. The United States also continued to echo the Bush administration's call for the inclusion of all carbon sources and sinks. The international community saw this approach as an attempt by the United States to claim its vast forests as a mechanism for reducing greenhouse gas emissions, thereby avoiding having to actually decrease emissions. Finally, the United States lobbied for limiting emissions from all countries, counter to their stance at COP-1, where an agreement was reached not to require reductions from developing countries.[23] In the end, more than one hundred countries announced that they would develop binding greenhouse gas reduction targets.

As the United States continued the dialogue on the international front, support at home was lacking. On July 25, 1997, anticipating the COP-3 negotiations in Kyoto, Japan, at the end of the year, the U.S. Senate passed the Byrd-Hagel resolution by a unanimous vote (95–0) declaring its opposition to any international agreement that either harmed the U.S. economy or did not require the participation of developing countries.[24] While it was nonbinding, it was a strong statement on the position of Congress regarding the agreement expected to come out of the meetings in Kyoto.

With wavering support at home, President Clinton sent Vice President Gore to Japan to advise the U.S. negotiators to be more flexible during the discussions at COP-3. Ultimately, the U.S. negotiators' efforts were successful, and on December 11, 1997, the United States joined more than 150 nations in signing what has since become known as the Kyoto Protocol. Although the participating nations had reached and signed an agreement, the Kyoto Protocol contained language that would delay its implementation and coming into force as a legally binding treaty until it had been ratified by at least 55 countries representing over 55% of the world's CO_2 emissions (which would not happen for another seven and a half years).

The treaty committed the industrialized nations to legally binding reductions in their emissions of the six most important greenhouse gases (described in chapter one). These reductions varied from country to country, but if successful would cut overall global greenhouse gas emissions to 5% below 1990 levels by the years 2008 to 2012. For example, the United States agreed to reductions of 7%, Japan to 6%, while Russia only had to hold emissions constant, and in many cases the developing countries could allow their emissions levels to increase.[25] The United States' agreement to a reduction was contingent upon the creation of emissions trading among industrialized countries (a program in which nations can buy and sell rights to release greenhouse gases, as is explained in more detail later in this chapter).[26] Emissions targets were also included for the developing world, but were voluntary rather than compulsory.[27]

COP-4 met in Buenos Aires, Argentina, in November of 1998. Most of the concepts agreed to at COP-3 still needed to be addressed, including emissions trading, compliance and enforcement, carbon sinks, and the joint implementation programs and clean development mechanisms.[28] A two-year deadline was set to finalize the carbon reduction mechanism. During COP-4, the United States formally announced its signing of the Kyoto Protocol.[29] However, reiterating their position highlighted by the Byrd-Hagel resolution, the U.S. Senate criticized the terms of the treaty and emphasized that it would not be ratified by the United States until more developing countries agreed to binding reductions. The Senate argued that excluding developing countries would stifle the U.S. economy, and that domestic industries that emitted large amounts of greenhouse gases would relocate to unregulated countries. Recognizing that the chances of the Senate ratifying the Kyoto Protocol were nonexistent, the Clinton administration never even submitted it for ratification.[30]

Following the official signing of the Protocol in COP-4, American corporate opposition to greenhouse gas emission reductions began developing. It was recognized that emissions trading was a long-term process, and would not be implemented in time for the United States to meet its targets. As such, absolute emissions reductions would be required. As an alternative, by COP-5, the United States significantly increased pressure to earn credit for carbon sinks from agricultural and forestry practices.[31] The European Union balked at the carbon sink strategy, and insisted that at least half of the industrialized countries' reduction targets be achieved through some domestic cuts in fossil fuel emissions.[32] Negotiations continued through these meetings and into COP-6, which also ended without an agreement.

In his final State of the Union address in January 2000, President Clinton expressed his continued concern about the severe impacts of global warming. He stated: "The greatest environmental challenge of the new century is global warming . . . If we fail to reduce the emissions of greenhouse gases, deadly heat waves and droughts will become more frequent, coastal areas will flood and economies will be disrupted. That is going to happen, unless we act . . . New technologies make it possible to cut harmful emissions and provide even more [economic] growth."[33]

In his final year in office, President Clinton proposed $2.4 billion in the fiscal year 2001 budget to combat global warming (though Congress only awarded 40%). The funding supported key initiatives to develop clean energy sources, and to spur state and local efforts to reduce air pollution (and the corresponding greenhouse gas emissions). In addition, his budget called for $1.7 billion for further research on both climate science and adaptation to climate change through the U.S. Global Change Research Program.[34] Furthermore, the administration made significant progress toward increasing energy efficiency appliance standards, which will result in saving consumers billions and preventing millions of tons of carbon dioxide from being emitted.[35]

Overall, the Clinton presidency was one of moderate U.S. advancements in combating global climate change. A number of new initiatives were implemented, and President Clinton did send Vice President Al Gore to assist with the negotiations at COP-3 in Kyoto. Even though the Kyoto Protocol did not go into effect in the United States, because ultimately the Senate did not ratify it, President Clinton made an important statement by signing the Protocol at COP-4. More than anything, the Clinton administration raised awareness of the significance of climate change as an important issue, worthy of a national debate.

The U.S. Position in the Early Twenty-First Century: The Bush Administration

During 2000, then Republican presidential candidate George W. Bush offered explicit support for the environment by promising to reduce carbon dioxide emissions as part of an overall strategy to reduce emissions from power plants. Yet three months after taking office, President Bush wrote that he was not willing to regulate carbon dioxide emissions (which are not listed as a pollutant under the Clean Air Act). President Bush argued that regulating CO_2 would force power generators to shift from coal to natural gas, and in light of the California energy crisis, consumers would be burdened. He also cited "the incomplete state of scientific knowledge of the causes of, and solutions to, global [climate] change and the lack of commercially available technologies for removing and storing carbon dioxide."[36] Two weeks later, President Bush rejected the United States' participation in the Kyoto Protocol, stating, "I will not accept a plan that will harm our economy and hurt

American workers."[37] He insisted that by adhering to the Kyoto Protocol's absolute emissions reductions mechanisms, the U.S. economy would be affected adversely. In doing so, he returned to the hard line of his father's presidency, which held that the United States would not take any action to reduce greenhouse gas emissions that would affect how business is done in this country.

It took another year for President Bush to release his climate change strategy. The Bush administration's Global Climate Change Initiatives aimed to cut greenhouse gas intensity (emissions per dollar of real gross domestic product) by 18% over ten years. Under this approach, emissions are reduced in relative terms, but the United States' overall emissions level can continue to grow as the economy expands. The target was to reduce the United States' annual domestic carbon emissions, through voluntary corporate action, from 183 to 151 metric tons per million dollars of gross domestic product by 2012.[38] The stated aim of this program was to "slow the growth of greenhouse gas emissions, and—as the science justifies—to stop, and then reverse that growth."[39]

The White House also continued to downplay the importance of global warming. One of the more egregious actions was to edit the section on global warming in the Environmental Protection Agency's "Report on the Environment." Areas regarding human contribution to climate change and the sharp temperature increases experienced over the past decade were removed, as well as the sentence, "Climate change has global consequences for human health and the environment."[40] Despite international acceptance for the research and findings of the IPCC, President Bush's strategy for a U.S. federal climate change policy was to focus millions of dollars on additional studies to research the threat posed by climate change. In July 2003, the Bush administration released the Strategic Plan for the Climate Change Science Program. Citing "the President's direction that climate change research activities be accelerated to provide the best possible scientific information," the Strategic Plan proposed further research in the following five areas:

1. improve knowledge of the Earth's past and present climate;
2. improve quantification of contributors to climate change;
3. reduce uncertainty in how the Earth's climate may change;
4. understand how humans and ecosystems will respond to climate change; and
5. explore how to manage the risks and opportunities of a changing climate.[41]

With lack of White House support for addressing the issue of global warming other than researching it further, two senators moved to include the United States in the battle against climate change. Joseph Lieberman (D-CT) and John McCain (R-AZ) sponsored a bipartisan bill aimed at curbing U.S. greenhouse gas emissions. The Lieberman-McCain Climate Stewardship Act of 2003 was presented to Congress as a chance to salvage U.S. climate change participation and to show a commitment to a healthier, cooler Earth. The groundbreaking legislation proposed a "cap and trade" system that would harness market forces to help cut greenhouse gas emissions. The Senate considered the bill on October 29–30, 2003. Ultimately, it was rejected—43 in favor and 55 opposed along a largely party-line vote—with only 6 Republicans voting in favor and 10 Democrats opposed.[42] The bill has been reintroduced in subsequent congressional sessions with little action taken.

Meanwhile, during the early years of the Bush administration, international action continued in the same vein as at the end of the Clinton administration. Additional countries continued to ratify the Kyoto Protocol, and the subsequent COP meetings (7, 8, 9, and 10) focused on continued negotiations and working out the implementation mechanisms.

In late 2004, a major breakthrough occurred when the Russian Parliament voted to adopt the Kyoto Protocol and its targets. This move brought the total percentage of greenhouse gas emissions from ratifying countries above the 55% threshold, and the Kyoto Protocol cleared the last major hurdle to becoming a legally binding instrument for all countries that had ratified it.[43] This was considered a major victory for the climate protection movement, as common thinking was that the Protocol would not achieve sufficient support without ratification by either the United States, Australia, or Russia, and the first two countries showed no intent of ratifying the Protocol. On February 16, 2005, the Kyoto Protocol took effect as a binding commitment for the countries that ratified it.

The year 2005 saw another breakthrough in climate negotiations, this time in Montreal where the first international meeting was taking place since the Kyoto Protocol went into force. At COP-11/MOP-1, the countries bound by the Kyoto Protocol established a process to negotiate further and deeper cuts after the 2012 expiration of the first commitment period of the Kyoto Protocol.[44] This decision was made despite significant opposition from the U.S. negotiators, whose credibility in the proceedings had dropped considerably with the Bush administration's withdrawal from Kyoto. Also at this meeting, the parties to the protocol adopted the "Marakesh Accords," the "rule book" outlining how the protocol is to be implemented. The joint implementation programs and the clean development mechanism methodologies were officially launched, and agreement was reached on the compliance regime, with enforcement and facilitative branches. Former President Clinton returned to the international climate change arena, addressing the delegates on the last day of the meeting and attacking the Bush administration's position against the Kyoto Protocol as "flat wrong." His message was that climate protection would benefit all sectors and strengthen the economy.

Recent Climate Change Policy Action

Although the Bush administration has been largely antagonistic to the idea of taking strong action on climate change, overtures have been made in Congress (largely unsuccessful) and at the state level since 2000.

Federal Activity

As mentioned previously, the most notable piece of legislation aimed at curbing greenhouse gas emissions in the United States was introduced by Senators McCain and Lieberman. Known as the Climate Stewardship Act, this bill was first introduced in 2003. If passed, it would have capped 2010 emissions of CO_2 at 2000 levels, required the formation of a national database of greenhouse gas emissions, and funded research on climate change. This bill was aimed primarily at industry and energy producers. Smaller sources from which it is harder to measure emissions, such as the residential and agricultural sectors, would be exempted. Although this bill ultimately failed, the margin of the vote (43 to 55) was closer than many expected and showed that support for action to protect the climate was growing.

Senators McCain and Lieberman reintroduced this bill in 2005 as the McCain-Lieberman Climate Stewardship and Innovation Act, which contained a greater emphasis on the development and deployment of new technology. It too ultimately failed, and by a larger margin (38 to 60), in part due to the addition of subsidies for construction of new nuclear power plants as part of the technology package. This addition caused four Democratic senators who supported the 2003 version of the bill to oppose the 2005 version.

Despite these failures, some pieces of legislation did make it out of Congress. In 2005, President Bush signed an energy bill containing an amendment inserted by Senator Chuck Hagel that provided incentives for the development of renewable energy technology. Although voluntary, this measure is particularly interesting in that Senator Hagel (R-NE) cosponsored the 1997 resolution against the Kyoto negotiations. Senator Hagel now says that climate change is a "top tier issue."[45]

In another act indicating that the political opinion was changing, the Senate also passed a nonbinding resolution in June of 2005. This "Sense of the Senate" resolution was introduced by Senator Bingaman (D-NM). It affirmed that there is scientific consensus that human emissions threaten the stability of the climate and that "Congress should enact a comprehensive and effective national program of mandatory, market-based limits and incentives on emissions of greenhouse gases that slow, stop, and reverse the growth of such emissions" in such a way that this action would not harm the U.S. economy and would encourage similar actions by other countries.[46] Although this amendment was stripped out of the final energy bill that was passed, it marked an important shift in the Senate—for the first time, a bipartisan majority recognized the need for a mandatory cap on greenhouse gas emissions.

Since that time, many additional bills have been submitted to Congress. Some key actions in recent years have included the following.

- Two bills were introduced (in the House and Senate) by Representative Henry Waxman and Senators Jim Jeffords (I-VT) and Barbara Boxer (D-CA) that would require an 80% reduction in greenhouse gas emissions below 1990 levels by 2050.

- The House approved a $1 million study to determine which coastal population centers in the United States bear the highest risk due to sea level rise caused by climate change.

- The Senate Foreign Relations Committee passed a resolution calling for the United States to reenter international climate negotiations.

- Senators Kerry (D-MA) and Snowe (R-ME) introduced legislation to freeze emissions in 2010 and reduce them to 65% below 2000 emissions levels by 2050. The bill includes provisions for a cap-and-trade program and requirements to reduce emissions from passenger vehicles, for 20% of the United States' electricity to come from renewable sources, for re-engagement in international climate negotiations, and for the establishment of a national program to assess the extent of community vulnerability to climate variability.

- According to the American Geological Institute, Congress held fifteen hearings on aspects of global climate change between January 2005 and December 2006.[47]

It remains to be seen which direction Congress and the federal government will take on this issue in the future. With the Democrats taking control of both houses of Congress in the 2006 elections, the 110th Congress will have significantly different leadership than has been seen in the preceding decade—the decade in which international consensus congealed around the need for strong action to be taken to combat global warming. Included in this leadership change, Senator Boxer will take over leadership of the Senate's Environment and Public Works Committee from Senator Inhofe (R-OK) (who has called climate change "the greatest hoax ever perpetrated on the American people").[48]

Additionally, the U.S. Supreme Court has declared, in a case on climate change brought forward by a coalition of twelve states and various environmental and public health groups, that the U.S. EPA does have the authority to regulate greenhouse gases. The suit claimed that they, and the American people, are being harmed by the EPA's failure to regulate greenhouse gas emissions. Up until this decision, the EPA (under the Bush administration) held that greenhouse gases were not classified as pollutants subject to regulation under the Clean Air Act. The Court ruled that "under the [Clean Air] Act's clear terms, EPA can avoid promulgating regulations only if it determines that greenhouse gases do not contribute to climate change or if it provides some reasonable explanation as to why it cannot or will not exercise its discretion to determine whether they do."[49]

Finally, in the lead-up to the 2007 G8 Summit in Germany, President Bush made the proposal that the United States and other nations should meet to develop a new framework for reducing greenhouse gas emissions that could be put into place by the end of 2008. Although the suggestion that the United States participate in an international climate discussion would appear to be a step forward, some critics claim this tactic to be a diversion to avoid the adoption of climate change prevention principles at the G8 Summit and to set up a parallel process that could undermine the Kyoto protocol.[50]

State Action

While the federal government has failed to take strong action on climate change, states have begun to address the issue. Since 1990, a number of states have conducted greenhouse gas inventories and some local emissions registries have been developed. California and the northeastern states have emerged as hotbeds of such activity.

In 2003, a cooperative of eleven New England and Mid-Atlantic states initiated the Regional Greenhouse Gas Initiative (RGGI). Together, they are taking a regional approach to reducing carbon dioxide emissions through creating a local cap-and-trade program. In 2006, RGGI released their model regulations for implementing this program, making this the first plan in the United States requiring mandatory emissions reductions. By 2009, the participating states will cap annual greenhouse gas emission from power plants at 121 million tons and then reduce those emissions by 10% before 2019.

California also has been a leader in tackling climate change. In 2002, the state adopted Assembly Bill 1493 (the "Pavley law"), which directs the state to develop regulations that reduce greenhouse gas emissions from passenger vehicles sold in California by the maximum extent feasible. This initiative is currently stalled in the courts but continues California's leadership role of cleaning the private vehicle fleet (California has long been a leader in reducing air pollutants from motor vehicles).

In 2006, California adopted another landmark piece of legislation. Assembly Bill 32 requires the state to put into place a cap-and-trade system for greenhouse gases from major emitters. The stated target of this bill is to reduce emissions to 1990 levels by 2020. It requires the state to adopt an appropriate cap by 2008, develop a plan for meeting the reduction target by 2009, and adopt regulations to achieve the maximum feasible (and cost-effective reductions) in greenhouse gas emissions by 2011.

Until the United States adopts credible climate policy on the federal level, any policy action addressing climate change in the United States has to happen voluntarily at the state or local level.[51] Since so many difficulties exist in the adoption of national or international greenhouse gas reduction initiatives, local governments in cities and counties around the country are implementing policies dealing with climate change and other national or international problems. Not only can these strategies be implemented rapidly with less bureaucracy at the local level than can federal policy initiatives, they can be customized to localized needs instead of uniformly imposed across the nation. Additionally, local governments begin to experience the many ancillary benefits to reducing global warming emissions at the local level, such as financial (taxpayer) savings, health benefits, and overall improved quality of life. Chapter Three begins to explore these results.

Market-Based Greenhouse Gas Regulation Options

To date, all major state, federal, and international climate change policy options that have been introduced rely on market-based approaches rather than traditional "command and control regulation."[a] Market-based and flexible approaches to environmental policy are also the preferred policy tool of environmental economists.[b] Under command and control systems, regulators impose national standards for emission of pollutants by industries and enforce the standards through active monitoring of factories and imposition of civil or criminal penalties for noncompliance.

Market-based approaches are incentive-based policies that attempt to encourage conservation practices or pollution reduction strategies rather than force polluters to follow a specific rule.[c] These are viewed by economists as less intrusive into the capitalistic system and are advocated by policymakers who prefer to avoid additional regulations that often are seen as an undue burden caused by the federal government.

The success of programs such as the Sulfur Dioxide (SO_2) Allowance Trading Program, which has helped reduce sulfur emissions (responsible for acid rain) by 50% (10 million tons from 1980 levels) without restrictive regulation, indicate that using market-based solutions to policy issues such as global warming may be an effective national or international approach.

President Clinton's proposed carbon tax was an example of a market-based mechanism as well. The theory is that by imposing a tax on the source of pollution (i.e., the fossil fuels that emit CO_2), prices for those goods will go up, which will in turn drive down demand and spur alternatives, limiting emissions. When applied for maximum impact, the funds from the tax also will be reinvested to mitigate global warming or to fund research and help bring to market alternative, non-carbon-based fuels and technologies. Thus, there are not only reduced emission levels from a decreased demand in fossil fuels, but also a source of funding to be reinvested into programs to help erode emission levels.

The market-based approaches incorporated into the Kyoto Protocol include emissions trading (cap-and-trade), the clean development mechanism, and joint implementation programs. Emissions trading is a method where an established baseline level of emissions is determined. If a country is emitting less than its assigned level of emissions, it can sell the remaining credits to a country that cannot reduce emissions below its assigned level. For example, country A is permitted to emit 100 tons, and country B 120 tons. If country A has effectively reduced emissions to 85 tons, it has a credit of 15 tons. If country B cannot reduce its emissions lower than 130 tons, it can buy 10 of country A's emission credits and still be considered in compliance.

The clean development mechanism is a tool to promote investments in greenhouse gas emissions reduction projects in developing countries, which are otherwise unlikely to be implemented. They are structures set up by the Kyoto Protocol to allow industrialized countries with a greenhouse gas reduction target to invest in and claim emissions reduction credits for projects carried out in developing countries that do not have Kyoto targets. This provides the industrialized countries with a mechanism to reduce the cost of meeting their commitments. For example, if it is cheaper for a country in the European Union to replace a fossil fuel power plant in an African nation than to carry out the same project at home, the clean development mechanism allows for this foreign investment to count toward the emissions reduction target of that European country.

Joint implementation programs are very similar to the clean development mechanism programs, except the exchange takes place between developed countries with established Kyoto targets. Taken together, these mechanisms allow countries to meet their individual emissions reduction targets at the least possible cost.

It is a safe assumption that U.S. climate policy will rely on similar market-based policies for reducing emissions. There is generally a feeling that market-based approaches to reducing emissions, which set limits and then let the emitters determine the most cost-effective way to meet those caps, are less intrusive than prescriptive regulations, which mandate specific actions. Similarly, these tactics have a proven track record in existing state and international climate policies and federal regulations on other air pollutants.

a. Scott J. Callan and Janet M. Thomas, *Environmental Economics and Management* (Chicago: Irwin, 1996).

b. Ibid.; Atle Christer Christiansen, "Convergence or Divergence? Status and Prospects of U.S. Climate Strategy," *Climate Policy* 3 (2003); and Michael E. Kraft, "Environmental Policy and Politics in the United States: Toward Environmental Sustainability?" in *Environmental Politics and Policy in Industrialized Countries*, ed. Uday Desai (Cambridge: MIT Press, 2002).

c. Callan and Thomas, *Environmental Economics*.

Chapter Three

Local Actions, Global Results

I want my grandchildren to be able to breathe good
clean air![1]

Mayor Kenneth E. Patton, Brooklyn, Ohio

W hile the United States government has hardly taken a leadership role
in curbing greenhouse gas emissions, cities and counties across the country
have become involved in a movement to address global warming at the local
level. Although climate change is a global issue and debates over the reduction of greenhouse gases occur primarily at the national and international
levels, local communities have realized that they are on the front lines in
terms of both mitigating and adapting to global warming. If nothing is done
to slow the rate of global climate change, it is local communities that ultimately must deal with its impacts in order to protect their citizenry.[2] Additionally, if and when action is taken to reduce emissions, local governments
will be largely responsible for carrying out those actions, as it is ultimately at
the local level that greenhouse gas emissions occur. City and county governments pass and enforce building and energy codes; collect and dispose of
waste; and develop and maintain transportation infrastructures, all of which
dramatically affect the amount of greenhouse gases released by the United
States.

Local officials have long taken a leadership role in recognizing and react-
ing to the potential dangers of changing global temperatures. Indeed, the
first governmental entity to adopt a greenhouse gas emissions reduction target was a city government. At the World Conference on the Changing Atmosphere: Implications for Global Security, held in the City of Toronto, a
statement was issued to the industrialized nations of the world encouraging
them to reduce their emissions to 20% below 1988 levels by the year 2005.
The City of Toronto led the way by officially adopting that target. This stood
as the only official emissions reductions target for many years and formed
the starting place for international talks that led to the adoption of the
UNFCCC in Rio in 1992 (although the target was reduced before adoption

and listed as voluntary). By the Second Municipal Leaders Summit on Climate Change held in March 1995 in Berlin, Germany, thirty cities had adopted the "Toronto Target" (two years before the Kyoto Protocol was agreed upon).[3] A 20% emissions reduction target remains the most commonly adopted target by local governments in the United States.

Municipal leaders have continued to be proactive in addressing global climate change at home and at higher levels of government. At the COP-11/MOP-1 meeting in Montreal, local leaders took part in the "Fourth Municipal Leaders Summit on Climate Change." At this four-day event, held simultaneously with the international negotiations, the 309 participants (representing thirty-eight countries) attended sessions and took part in discussions ranging from "responding to the challenges of climate change" to "municipal movements and best practices" and the links between "local climate action and sustainability." This summit culminated with the local leaders in attendance passing a resolution that was read on the floor of the main COP/MOP plenary session. For many of the leaders present, this resolution reiterated the commitments they have held for more than fifteen years. The declaration:

- recognizes climate change as an issue of importance to local communities that local governments have the ability to affect;

- commits to reducing emissions locally through innovative actions, monitoring of emissions levels, and development of strategic partnerships to enhance the jurisdictions' reduction potentials; and

- requests that local governments be included in the international negotiations, polices, and trading mechanisms.[4]

During the summer of 2005, the critical role that local action could play in curbing emissions came to the forefront of the climate debate in the United States. On February 16, 2005 (the day the Kyoto Protocol went into effect), Mayor Greg Nickels of Seattle challenged U.S. cities to join Seattle in meeting or exceeding Kyoto Protocol guidelines. The U.S. Conference of Mayors passed a resolution endorsing the U.S. Mayor's Climate Protection Agreement that Mayor Nickels drafted earlier that year and has been signed by 358 U.S. mayors in forty-nine states (as of late 2006). These communities represent over 55 million citizens.[5] This agreement commits communities to:

- Strive to meet or beat the Kyoto Protocol targets in their own communities, through actions ranging from anti-sprawl land-use policies to urban forest restoration projects to public information campaigns;

- urge their state governments and the federal government to enact policies and programs to meet or beat the greenhouse gas emission reduction target suggested for the United States in the Kyoto Protocol—7% reduction from 1990 levels by 2012; and

- urge the U.S. Congress to pass the bipartisan greenhouse gas reduction legislation, which would establish a national emission trading system.[6]

Beyond simply passing resolutions, local governments are educating themselves on what can be done and taking action. In July 2005, forty-five U.S. mayors came together in Sundance, Utah, to participate in a climate change summit. There they took part in an intensive series of educational forums on the science of climate change, mitigation technologies, and economic and business tools. Keynote speakers included scientific experts, local political leaders, former Vice President Al Gore, and conference host Robert Redford.[7] This unique gathering of mayors is one in a long series of events in which elected officials and staff members have taken part to reduce emissions.[8]

Local Government Strategies

Although multiple actions promoting climate protection at the local level have come to prominence recently, the concern over global warming on the part of towns, cities, and counties is not new. Rather, it is a movement that has been growing for over a decade, largely cultivated by the international local government association ICLEI—Local Governments for Sustainability (commonly known as ICLEI).[9] Founded in 1990 at the United Nations Congress of Cities for a Sustainable Future, ICLEI is an international nonprofit association of local governments dedicated to addressing environmental problems through cumulative local actions. Currently more than 650 local governments from around the world participate in the campaigns and programs offered by ICLEI to implement innovative environmental management procedures at the local level.[10] ICLEI provides these local agencies with training, tools and technical assistance, and informational resources. It also acts as an international clearinghouse for best practices and as a developer of a comprehensive framework to address regional and global environmental problems at the local level.

The Cities for Climate Protection (CCP) Campaign is ICLEI's flagship program. Initiated with fourteen pilot cities in January 1993 at the First Municipal Leaders Summit on Climate Change in New York, the CCP Campaign has now grown to work with over 237 cities and counties in the United States representing approximately 20% of the U.S. population.[11] The campaign's goal is to reduce greenhouse gas emissions resulting from the burning of fossil fuels and other anthropogenic activities that contribute to global warming and air pollution. Actions that reduce these emissions not only protect the global climate but also improve the quality of life in local communities. Any city or county can join the CCP by passing a legislative resolution committing to take steps to quantify and reduce their greenhouse gas emissions.

Participants in the CCP commit to reducing local emissions that contribute to global warming by working through the CCP's 5-Milestone process:

1. Conduct a local emissions inventory.
2. Adopt an emissions reduction target.
3. Draft an action plan to achieve the target.
4. Implement the action plan.
5. Evaluate and report on progress.

Cities across the country, ranging in population from under 2,000 to over 8 million, have pledged to take leadership roles in curbing greenhouse gas emissions generated at local levels. On their own, each city's battle to reduce emissions may not make a significant difference, but as part of a growing campaign involving numerous jurisdictions, they have incredible potential to stabilize atmospheric levels of greenhouse gases.

ICLEI provides jurisdictions with tools and training for quantifying emissions levels, setting targets, and measuring the impacts of emission reduction measures. These programs focus on the co-benefits of emissions reduction as well, such as the impacts of action on the regulated "criteria air pollutants," economic pay-back periods, and more. ICLEI also coordinates regional networks of jurisdictions so that communities can build region-wide strategies that magnify the impacts of their programs. Additionally, through program-specific assistance, ICLEI helps local governments implement tangible emission reduction programs in various sectors of their communities, such as business, transportation/fleet, land-use, or waste.[12]

Climate Leadership—The Case of Seattle

One cannot talk about local governments' efforts to reduce greenhouse gas emissions without mentioning Seattle, Washington. Seattle was one of the original fourteen local governments to participate in ICLEI's Cities for Climate Protection campaign and remains an international leader in local action. Mayor Greg Nickels is a staunch advocate of reducing greenhouse gas emissions, and has led a vocal and public push for more municipal recognition and response toward global warming at the local level. In February 2005, Mayor Nickels challenged cities to join Seattle in meeting or exceeding Kyoto Protocol guidelines, and on June 13, the U.S. Conference of Mayors passed a unanimous resolution, sponsored by Mayor Nickels, that urged Congress to recognize global warming as a human-induced threat and to enact the Kyoto policies.[a]

To guide Seattle's march toward the Kyoto Protocol goals, Mayor Nickels created a "Green Ribbon" commission of local leaders charged with developing a climate action plan for the city. In early 2006, the task force presented a plan with specific attainable emissions goals to:

- reduce automobile dependence (170,000 tons);

- increase fuel efficiency and the use of biofuels (200,000 tons);

- achieve efficient homes and businesses that use clean energy (316,000 tons);

- continue to build on the city's leadership role;

- sustain the city's commitment into the future.[b]

The City of Seattle has a long history of environmental activism; late in the 1970s, it was the only city that rejected nuclear power and chose instead to invest in energy conservation. The success of energy efficiency as the preferred alternative led the city to invest in recycling instead of an incinerator to manage solid waste, and to choose water conservation over developing a new source of supply to serve the growing population of greater Seattle. Seattle also committed to reducing urban sprawl and offering transit alternatives, including light rail, buses, streetcars, and improved bike and pedestrian opportunities. Seattle City Light is the only electric utility in the country that has achieved zero net greenhouse gas emissions, having adopted an aggressive strategy for investing in projects to offset the CO_2 emissions associated with all power purchases.[c] Internally, emissions from municipal operations have been reduced almost 60% below 1990 levels, and Seattle is working toward a goal of zero emissions.[d] Through these actions, Seattle has been working toward reducing the city's contribution to climate change—and many of Seattle's climate protection activities preceded participation in CCP.

a. J. M. McCord, "Seattle Leads Cities in Reducing Greenhouse Gases," *Aspen Weekly*, July 23, 2005.
b. Mayor Nickels' Green Ribbon Commission on Climate Protection, *Seattle, a Climate of Change: Meeting the Kyoto Challenge*, Report and Recommendations, 2006.
c. "City Light First in Nation to Reach Zero Net Emissions Goal," press release, www.seattle.gov/news/detail.asp?id=5656&dept =40 (accessed on February 18, 2007).
d. McCord, "Seattle Leads"; and www.seattle.gov/environment/ climate_protection.htm (accessed on May 1, 2006).

Results from the Local Level

How much of a difference can municipalities make? Often there is an assumption that local governments are too small and localized to have a dramatic impact on overall greenhouse gas emissions. But taken as a whole, cities and counties each doing what they can at home could result in a significant effect. For example, the 650 jurisdictions around the world that participate in ICLEI's CCP campaign account for over 15% of global greenhouse gas emissions. The most recent survey of emission reduction activities by local communities in the United States (released in 2005) demonstrated impressive results. Actions and programs undertaken by these local governments have:

- prevented the release of *23 million tons* of eCO_2 emissions;

- reduced electricity consumption by *4,000 gWh*;

- saved *74 million gallons* of transportation fuel (gasoline and diesel); and

- saved *6 million therms* of natural gas.[13]

These are conservative estimates of the emissions reductions coming from action by local governments. This quantification accounts only for local governments that participate in the CCP campaign that responded to a self-assessment questionnaire of programs that they could quantify relatively easily the emission reductions. In reality, the effect of local governments involved in the CCP is likely to be double what is reported here, and the impacts of all local government action in the country is likely to be many times this number. (As there is no more-comprehensive quantification process for quantifying and reporting greenhouse gas emissions reductions, this number is reported here as an indictor of the level of action taking place at the local level.)

Overall, local governments can and are having an impact on global emissions levels. Although these reported reductions only lower the total U.S. emissions by 0.2 percent (in 2004, the United States emitted 7,799 million tons of eCO_2), they are impressive considering that there are no regulations compelling local action.[14] A 23-million-ton reduction in greenhouse gas emissions is roughly equivalent to eliminating the emissions from 1.8 million households, or removing 4 million passenger from the road for a year. As more cities begin to participate in this process of combating global warming, and as the current players increase their efforts, there stands to be an even greater decrease in the levels of eCO_2 entering the atmosphere.

Local Action—More than Combating Global Warming

While these reductions in greenhouse gas emissions are impressive, local governments are realizing many additional co-benefits. Regional air pollution and global climate change are separate environmental problems, but their causes and solutions are closely linked. Recognizing the multiple benefits of reduced emissions can increase willingness of governments and their citizenry to participate in such programs. By crafting smart policies that recognize the co-benefits of actions that reduce both air pollution and greenhouse gas emissions, and also save money, state and local governments can achieve a wide range of environmental, public health, and economic goals efficiently.

Cleaner Air and Reduced Health Problems

When climate-altering gases are emitted during fossil fuel combustion, many other harmful gases are also released, all of which prove detrimental to citizen health and quality of life. The American Lung Association affirms that air pollution from fossil fuel sources contributes to lung diseases, such as respiratory tract infections, asthma, and lung cancer. Lung disease claims close to 335,000 lives in America every year and is the third-leading cause of death in the United States. Over the last decade, the death rate for lung disease has risen faster than that of any of the top five causes of death.[15] The most common air pollutants, classified as the criteria air pollutants by the EPA, are nitrous oxides (NO_x), sulfur dioxides (SO_x), carbon monoxide (CO), volatile organic compounds (VOC)—a precursor to the formation of ozone—and particulate matter (PM10). Actions put into place by local governments to reduce their greenhouse gas emissions are also reducing the amount of these air pollutants released by 43,000 tons annually.[16]

Nitrogen Oxides. Nitrogen oxides (NO_x) comprise a family of gases that are made up of nitrogen and oxygen molecules. This family of gases contains various nitrates, nitric oxide, and nitrogen dioxide. They typically are released into the atmosphere during the combustion of fossil fuels, primarily from automobiles and power plants. Nitrogen dioxide is a brownish gas that contributes to the formation of ground-level ozone (a principal component of smog) and nitric acid (a component of acid rain).[17]

HEALTH EFFECTS

According to the American Lung Association, nitrogen dioxide can irritate the lungs and lower resistance to respiratory infections such as influenza. While the effects of short-term exposure remain uncertain, exposure to higher concentrations can cause an increased frequency of acute respiratory disease in children.[18]

NO_x also can react with other compounds to further impair human and ecosystem health. Ground-level ozone, or smog, is formed when NO_x and volatile organic compounds (see below) react in the presence of heat and sunlight. Children, the elderly, people with lung diseases such as asthma, and people who work or exercise outside are especially prone to adverse effects such as damage to lung tissue and reduction in lung function.

NO_x and sulfur dioxide (see below) also react with other substances in the air to form acid rain. Acid rain can damage buildings, statues, and monuments, deteriorate cars, and cause lakes and rivers to become acidic and unsuitable for aquatic life. Human health concerns include effects on breathing and damage to lung tissue. Small particles penetrate deeply into sensitive parts of the lungs and can cause or worsen respiratory disease such as emphysema and bronchitis, and aggravate existing heart disease.[19] Furthermore, wind currents transport ozone and acid rain significant distances, allowing these pollutants to affect regions hundreds of miles downwind from the original source.

Sulfur Dioxide. Sulfur dioxide (SO_2) is one of many gases classified as a sulfur oxide gas (SO_x). These gases form when sulfur-containing fuels such as coal are combusted, and as a byproduct of industrial processes. The American Lung Association reports that the highest concentrations of SO_2 occur within the vicinity of large industrial facilities.[20]

HEALTH EFFECTS

High concentrations of SO_2 negatively affect breathing and can lead to respiratory illness and aggravation of existing cardiovascular disease. Those especially sensitive to SO_2 are people who already have impaired lung function (such as asthma or other chronic lung diseases). Children and the elderly are also at risk.[21] Sulfate particles are the major cause of haze and smog in many parts of the country, including our national parks.[22] As mentioned, SO$_x$ compounds also contribute to the formation of acid rain. The EPA notes that SO$_2$ damages forests and crops and acidifies soils, lakes, and streams. Continued exposure alters ecosystems by affecting the natural variety of plants and animals. The built environment is not immune to the negative effects of SO$_2$ either—the acid rain it produces causes buildings and monuments to corrode at an accelerated rate.[23]

Carbon Monoxide. Carbon monoxide (CO) is a colorless, odorless, and at higher levels, poisonous gas formed when carbon in fuel is not burned completely. The majority (60%) of carbon monoxide emissions are from vehicle exhaust, and thus the highest concentrations occur in areas of heavy traffic congestion. The American Lung Association states that in urban areas, motor vehicles account for as much as 95% of all CO emissions, although a host of other sources exist, such as industrial processes, nontransportation fuel combustion, and natural sources such as wildfire.[24]

HEALTH EFFECTS

When carbon monoxide is inhaled, it begins to reduce oxygen delivery to the body's organs and tissues.[25] Those with the highest risk from lower levels of CO are people suffering from heart disease. A single exposure may cause chest pain and reduce that person's ability to exercise; repeated exposures may contribute to other cardiovascular problems.[26] Even healthy people can be affected by elevated levels of CO. Those who breathe high levels of CO can become subject to a variety of ailments, including vision problems, reduced manual dexterity, and difficulty performing complex tasks. CO quickly becomes lethal at extremely high levels.[27]

Volatile Organic Compounds. Volatile organic compounds (VOC) are a principal component in atmospheric reactions that form ozone and other photochemical oxidants (commonly referred to as smog). VOCs are emitted from diverse sources, including automobiles, chemical manufacturing facilities, drycleaners, paint shops, and other commercial and residential sources that use solvent and paint. The term "volatile organic compound" is defined in federal rules as a chemical that participates in forming ozone.[28]

HEALTH EFFECTS

VOCs can cause a variety of health problems, ranging from minor eye, nose, and throat irritation and headaches, to damage to the liver, kidney, and central nervous system. Some compounds have been shown to cause cancer in animals and others are suspected or known to cause cancer in humans. The known health effects of VOCs vary greatly, ranging from those that are highly toxic to others that have no known human ramifications.[29]

Particulate Matter. Though not a gas, particulate matter can be especially dangerous when inhaled. Originating from a variety of sources, from smokestacks to construction activities to vehicle emissions, fine particles less than 10 microns (PM_{10}) can evade the respiratory system's natural defenses and enter the lungs.[30] Particles less then 2.5 microns ($PM_{2.5}$) are even more elusive to human defenses, as they can be inhaled deeply into the lungs and absorbed into the bloodstream or remain embedded for long periods of time. According to the American Lung Association, mortality risk increases 17% in areas with higher concentrations of small particles.[31]

HEALTH EFFECTS

People with impaired lung function are especially at risk to particulate matter air pollution. Effects include lung disease such as asthma, chronic bronchitis, and emphysema. Exposure can trigger asthma attacks and cause wheezing, coughing, and respiratory irritation and potentially lead to premature death from respiratory illnesses, cardiovascular disease, and cancer.[32]

• • •

Fortunately, when local governments take action to reduce their contribution to global warming, in most instances, they also reduce the emissions of other air pollutants that negatively affect the health of their citizens. These additional health benefits then serve as an added incentive to address climate change proactively. A well-thought-out greenhouse gas emissions reduction plan can have a significant impact on standard air pollutants. Table 3 illustrates the amount of pollutants that are prevented by actions that CCP participants took to reduce greenhouse gas emissions.

Table 3. **Estimated Air Pollution Prevention
by CCP Participants**

Air Pollutant	Pounds Reduced
NO_x	2,322,955
SO_x	35,880,909
CO	53,809,959
VOC	5,609,897
PM_{10}	6,304,146

SOURCE: Numbers are from ICLEI's quantification databases, queried spring
2005 (quantification methodology is explained in Appendix E).

Economic Benefits

Another complementary benefit of implementing measures to reduce
greenhouse gas emissions through energy efficiency is the financial savings
that accompany them. Decreasing energy consumption not only lowers fossil fuel use, but also leads to a corresponding decrease in energy costs for
households, businesses, organizations, and governments. Local governments
have numerous opportunities to incorporate energy efficiency into their procurement programs, many of which will be discussed in depth in the next
two chapters. Some common examples include switching to light-emitting-
diode (LED) stoplights, hybrid vehicle fleets, and EnergyStar office equipment. Over their lifecycles, these measures can yield significant financial
savings through lower energy bills, delayed maintenance, and longer lifetimes. In fact, CCP cities and counties are cumulatively saving in excess of
$535,000,000 in taxpayer money annually through reduced energy and fuel
costs.[33] Fiscal responsibility and the corresponding savings realized from
using the most-efficient, least-emitting technologies are powerful incentives
for any local government to take steps to reduce their emissions.

A prominent example of this is the EPA's EnergyStar® Buildings partnership program, with over two hundred state and local agencies participating. This program helps individuals, businesses, and governments track
their energy use and incorporate energy-efficient alternatives. Commercial
partners save more than $4.2 billion annually, while preventing 13.2 million
tons of eCO_2 from being emitted.[34] On the residential side, EnergyStar

products such as home electronics reduced energy consumption in 2004 by about 24.9 billion kilowatt-hours, avoided more than 13 million tons of carbon emissions, and saved consumers more than $5.1 billion. Overall, EnergyStar partners saved $9.7 billion and prevented 56.6 million tons eCO_2 in 2004.[35]

Recycling provides another opportunity for a community to save money and reduce eCO_2 emissions. Lower material costs and disposal fees are typical areas of cost savings. The more waste that is recycled, the lower the landfill fees associated with waste disposal. Boosting recycling rates is an effective strategy for reducing greenhouse gases in two different ways.

First, trash decomposition in landfills generates large quantities of methane. As Chapter One detailed, methane has over twenty times the heat-trapping potential as carbon dioxide. Thus, landfills across the country generate vast amounts of greenhouse gases each year. With recycling, trash is diverted from the landfill and put back into useful life. Lowering the levels of waste destined for a landfill results in lower levels of methane generation, significantly reducing greenhouse gas generation.

The Lifecycle of an Aluminum Can— The Case for Recycling

In the beginning, there was just bauxite (aluminum ore), probably from either the Australian outback or Jamaica. Bauxite is often found in extremely weathered rocks.[a] Before mining, at least the top six inches of top-soil must be removed from the site. Because bauxite is found near the surface of the ground (usually less than 100 feet), it is often mined by simple open-cast methods. Bauxite mining destroys more surface area than mining any other ore. From bauxite, alumina (Al_2O_3) is still most commonly extracted using a method that was developed back in 1888.

The amount of alumina produced is approximately half the weight of the original bauxite. A byproduct of caustic soda is captured for reuse and another toxic byproduct known as "red mud" usually is put into a nearby pond where some of it inevitably leaks into ground water. Next, the alumina is shipped across the world for further processing, using tons of energy and polluting the oceans.

Then the alumina is smelted. Smelting is very energy intensive: Making a single soda can of smelted aluminum is equivalent to burning a quarter-can (roughly 3 oz.) of gasoline.[b] In the smelting process, the alumina is dissolved in huge pots filled with cryolite (sodium aluminum fluoride) while carbon electrodes are added to the pot to send a giant 100,000 amps of electricity. In the process, carbon dioxide is produced, along with perflourocarbons (PFCs).

PFCs, which are greenhouse gases, trap heat like no other gases. PFCs are among the most harmful greenhouse gases to exist. The global warming potential (GWP) for perflourocarbons is 6,500 to about 9,200, compared to carbon dioxide with a GWP of 1. (GWP is the ability of a greenhouse gas to trap heat in the atmosphere relative to an equal amount of carbon dioxide.) Smelting is therefore one of the most destructive processes to the climate.

The next step in the process is shipping or trucking the aluminum slabs to a factory where the aluminum is flattened and then shipped to the next mill. At this mill, the aluminum is shaped into a can and printed with a design. The can is then baked twice and sent to the next factory. The can itself costs more to manufacture than the soda that will fill it.

---

a. Text excerpted from the University of California at Santa Barbara, "Life of a Soda Can," http://www.as.ucsb.edu/asr/Sodacan1.html (accessed on December 30, 2006). Research originally presented in John C. Ryan and Alan Thein Durning, *Stuff: The Secret Lives of Everyday Things* (Seattle: Northwest Environment, 1997).

b. Sources differ on the exact amount of gasoline it takes to produce an aluminum can, but in general the volume of gasoline used is reported as between π and 2/3 of a 12 oz. aluminum soda can.

The second contribution that recycling makes to reducing greenhouse gases is in reducing the indirect emissions associated with products. For example, a typical aluminum can begins as bauxite mined in Australia, smelted in China, and finally pressed into cans and filled in the United States. By using recycled cans, the whole process is completed domestically, reducing the emissions associated with mining new ore, smelting, and transworld shipping by as much as 75%.

Improving Quality of Life for Residents

Local governments that are taking action to reduce their greenhouse gas emissions are concurrently improving the quality of life for their residents. As mentioned, local governments in the Cities for Climate Protection (CCP) campaign have an additional $535 million dollars to reinvest in their local economies rather than paying electricity and fuel bills. This money can be used to improve schools, repair roads, increase city services, pay down debt, and to otherwise better the local community.

A secondary financial benefit of emission reduction activities is that these programs tend to shift the flow of money from sources outside the community to businesses within the local economy. Usually, utilities are not located within a municipality, meaning that money paid for energy purchases is leaving the local economy. Approximately $0.80 of every dollar spent on energy leaves the local economy. Alternatively, every dollar spent on renewables and efficiency has a 1.5 times greater impact on the local economy when compared to a dollar spent on fossil-fuel–based energy. This is because efficiency programs and use of renewable and alternative energy relies on local resources and local production capacity to a much greater extent than purchasing fossil-fuel–based energy (which is generated well outside of most communities). Therefore, these programs encourage spending in the local economy and generate additional local revenue.[36]

By decreasing greenhouse gas emissions, local governments are improving the health of their citizens. The significant amount of criteria air pollutants reduced by municipalities through local climate protection strategies provides a benefit to the community, nation, and world at large. Since air pollution can travel hundreds or even thousands of miles, citizens in the Midwest can benefit from initiatives implemented in California.

Decreased traffic congestion can be another co-benefit of reducing greenhouse gas emissions. Areas that offer mass transit options can reduce greenhouse gas emissions from automobiles while providing residents a cheap and quick method of transport around town. Recent studies have linked time commuters sit in traffic to billions of dollars of productivity.[37] Mass transit and altered land-use patterns also can improve mobility options for those who cannot drive or do not own a car. Many local governments are working to improve the walkability and bikeability of their jurisdictions. This allows bikers and pedestrians to travel in a safer environment. It also encourages citizens to walk or bike, which not only reduces congestion, but increases citizen health and well-being.

There are a host of reasons to implement programs to reduce greenhouse gas emissions besides reducing contributions to climate change. Local governments can save money, increase citizen health, and realize an improved quality of life for residents. In jurisdictions where local governments do not see climate change as a pressing issue, promoting the co-benefits can lead to successful implementation. And there are many options local governments can pursue to realize these benefits and reduce their contribution to global warming. Chapters Five and Six provide more detail on how local governments can begin to reduce their greenhouse gas emissions while saving money, improving air quality, reducing traffic congestion, and improving the quality of life for residents. But first, Chapter Four discusses the obstacles that must be overcome for local governments to begin to implement climate change policies successfully.

Chapter Four

Constraints on Local Policymakers

> Our number one priority is quality of life for our citizens—locally, nationally, and globally. I am pleased to join mayors across the country to work for a better environment. It is critical that cities be a major part of the solution, as we are a large part of the problem.[1]
>
> Mayor Mike McKinnon, Lynnwood, Washington

If there are so many benefits to reducing greenhouse gas emissions, why doesn't every municipality undertake these initiatives? Unfortunately, instigating significant change within a government structure can be a taxing process, and with far-reaching issues such as global climate change, local governments face multiple challenges. A fundamental barrier to action is getting an issue that is typically seen as a national or international issue onto the local agenda. Additionally, as Michelle Betsill identifies in her discussion paper "Localizing Global Climate Change: Controlling Greenhouse Gas Emissions in U.S. Cities," three additional barriers that local governments face include the government's own internal structure, staff availability to oversee a new program, and the availability of funds to finance action.[2] Municipalities may have to overcome one or a combination of these barriers, all of which can hinder the successful advancement of local climate policy.

Making a Global Issue Local— Getting It on the Agenda

Presentation is key to whether an issue will ever reach policymakers' attention. It is often a struggle to position global warming on local agendas because it is usually represented as a global issue requiring international solutions. This can leave municipal policymakers with the impression that local action will not affect the issue and prevent it from becoming a priority.

While most climate change studies focus on the national and international level, this information must be transferred to local policymakers in a language they can understand in order for successful local action to occur.[3] Additionally, the issue must be tied to local results. As Chapter Three identified, a host of co-benefits can result from addressing climate change on the local level. As these are tangible and resonate with local policymakers and their constituency, emphasizing such ancillary benefits, or even presenting them as the primary driver for action, is more likely to prompt changes in policy.

A local official who already has demonstrated concern for environmental issues can be an effective catalyst for change. Such policymakers are more likely to be moved to take action on climate change policies and to take advantage of the multitude of programmatic options available. As Betsill demonstrated in her research, mayors with interest in renewable energy, smart growth, and other "green" practices are more apt to "localize" global warming than mayors without prior environmental concern.[4] Furthermore, city staff are more inclined to find solutions to such challenging problems as local climate change policy when a mandate comes from the top.[5]

The lack of an appropriate "policy window" to introduce the topic can further hinder implementation of an emissions reduction campaign. In numerous communities, policymakers and their constituents do not see global warming as a pressing issue. Global warming is a topic not normally discussed by average citizens, and public information is sometimes inaccurate or out of date.[6] Indeed, a 1997 survey conducted by the Gallup Organization showed that the percentage of people that personally worry about global warming dropped from 35% in 1989 to 24% in 1997.[7] On the other hand, this apathy might have started to change in recent years. A 2005 Fox News poll found that 88% of Americans were aware of global climate change as a problem (as measured by respondents who classified climate change as a minor problem, a major problem, or as a crisis).[8] This shift in public perception could reflect the recent publicity that climate change has received through such high-profile events as the film *The Day After Tomorrow*, Hurricane Katrina, the Kyoto Protocol coming into force, the U.S. Conference of Mayors Agreement, and Al Gore's movie *An Inconvenient Truth*.

Despite increasing interest in the topic, the general population is still unclear on what the country can do about global warming, or whether action taken will even be effective. As a result, people may be upset about the problem, but their concern often translates into frustration rather than support for action.[9] Policymakers will not feel driven to action until their constituents have reached consensus on the importance of global warming and begin pressuring their leaders to take action. This highlights the need for continued public education and outreach about climate change. Similarly, policymakers who understand the issues have the double mandate to take action at the local level and to publicize the positive outcomes those actions will have for the community and the individual, such as those outlined in Chapter Three.

Another potential obstacle to translating scientific concern into public consensus and tangible action is the way that issues surrounding climate change are presented in the mass media. There is a high level of scientific agreement that global warming is an issue of concern. A 2004 study showed that of 924 articles on climate change in peer-reviewed scientific journals, none (0%) disagreed that human-caused climate change was occurring.[10] By comparison, a similar study of articles in the mass media showed that 53% of those articles reviewed gave equal time to climate change supporters and skeptics alike.[11] This discrepancy between what is being presented in the scientific community and what is being presented to the public can undermine the consensus needed by policymakers in order for them to take strong action.

Governmental Structure

Just as greenhouse gases originate from a diverse variety of sources, emissions reduction initiatives (like many other environmental programs) involve many departments within the government. In this case, the governmental structure itself can inhibit the initiation of an effective municipal program on climate change. Most city governments are divided into specialized departments, creating a "silo" effect that restricts communication and hinders the ability for a cross-functional approach to local initiatives.[12] As Betsill points out, effective climate change policy calls for collaboration from departments as diverse as waste management, transportation, public works, public utilities, public health, planning, and air quality management.[13] Cities without a coordinating entity, such as a department of the environment and sustainability or a climate and energy task force, face a larger obstacle to creating a holistic response that spans the jurisdictions of multiple departments.

Staff Availability

A third barrier to municipal action that Betsill identified is the staff time necessary to develop local policies and programs to control greenhouse gas emissions. Due in part to the barrier of bureaucratic structuring, addressing climate change at the municipal level is extremely time consuming. Simply to place that burden on staff already encumbered with other responsibilities is counterproductive to creating a successful climate change program—especially when existing issues originally have had a higher priority.[14] The ideal solution is to create a position or department specifically assigned to work on a local climate change program.[15]

Some risk is associated with where that position is located within the governmental structure. If it is part of the mayor's office, political turnover could cause the position or program to be eliminated. A better solution is to create an independent position that is less reliant on the current political climate within the government. However, failure to establish such an independent position or department is often due to the final institutional barrier—the availability of financial capital.

Fiscal Availability

While creating a position to address climate change is a fixed cost, to realize actual emission reductions, local governments will incur the additional costs of implementing new policies and programs. Unless the government increases revenues (through taxes) or eliminates or cuts back existing initiatives, climate change programs may have a difficult time finding funding from limited municipal budgets. Even if they are funded, these programs are often the first cut if they do not have strong political backing.[16] In times of budget deficits, money tends to be directed toward providing "traditional" services, such as maintaining schools and roads, rather than toward environmental problems that involve less tangible, longer-term results.

City budget officials are often conservative when investing in new programs without proven results, and can be hesitant to invest significant amounts of taxpayer dollars without a quantifiable return on investment involving the life-cycle costs of the projects.[17] Indeed, programs to reduce greenhouse gas emissions normally entail significant upfront costs. If these costs cannot be justified, with either a reasonable payback period or a significant reduction of greenhouse gas emissions, political support for the programs will decline.[18]

Despite these barriers, the local level still proves to be a setting highly conducive to implementing policies and initiatives that reduce greenhouse gas emissions. Chapter Five and Chapter Six describe actions that municipal governments can take to reduce their impact on global warming while saving money through efficiency and increasing the quality of life for their residents. The measures described fall into two major categories: community-based (Chapter Five) and local government-based (Chapter Six). Community measures address such sectors as residential buildings, commercial buildings, business practices, transportation, and waste management. Governmental measures pertain to reducing emissions from the government's own internal operations, including facilities, vehicle fleets, street and traffic lighting, employee commute, and the waste generated by municipal operations. Innovative examples of practices already implemented by cities around the country are provided for many of the measures.

Chapter Five

Global Issues, Local Action

Increasingly, cities are providing the answers to some of America's toughest problems. So it's fitting that we're leading the way on global warming as well.[1]

Mayor Dave Cieslewicz, Madison, Wisconsin

In spite of the barriers described in Chapter Four, local governments can be ideal places to address global warming for three related reasons. First, local governments adopt their own distinctive policies, appropriate for unique local circumstances. Local governments control many of the factors related to greenhouse gas emissions, such as energy codes, land-use decisions, residential and commercial regulations, transit options, and solid waste disposal.[2] Second, local authorities can encourage action by others in response to climate change, by lobbying the national government and by demonstrating the best-practices costs and benefits of controlling greenhouse gas emissions. Finally, municipalities have considerable experience addressing local environmental impacts within the fields of energy and waste management, transportation, and planning and development.[3]

Additionally, actions taken locally are a form of bottom-up environmental protection and resource conservation, and can act as microcosms for potential national policies.[4] Demonstrations of the effectiveness of mitigation options at the local level could make it more feasible for higher levels of government to adopt similar policies, and could make international actions more attractive. Research has indicated that there is a history of local governments demonstrating the effectiveness of policies, which are then adopted at higher levels of government.[5]

Local governments have a variety of opportunities to influence the emissions being released from their communities. Although the government is limited in its ability to take direct action to reduce emissions from the private sector, it can use policies, incentives, and investments to encourage emission reductions. By taking a leadership role in setting the framework for greenhouse gas reduction programs, local governments can see an increased participation level from all sectors and thereby realize a higher success rate in meeting their emission reduction goals.

Greening Local Buildings and Improving Efficiency

According to the U.S. Green Building Council, buildings consume 30% of all energy in the United States, and energy used in buildings accounts for 35% of U.S. CO_2 emissions.[6] Since buildings are responsible for such a large portion of U.S. energy use and greenhouse gas emissions, energy conservation in this sector can have a substantial effect on climate change and offer financial savings in the form of lower utility bills. Additionally, energy conservation measures, such as increased use of natural lighting and ventilation, better insulation, and efficient lighting, can improve comfort and indoor environmental quality. Improving working conditions in turn can increase productivity and improve the health of occupants (which also leads to financial savings for employers and individuals).[7]

Traditionally, local governments influence building energy use through the adoption of building codes, which can include prescriptive regulations that require design aspects that influence energy use (e.g., insulation, lighting), or standards-based regulations that mandate that buildings meet certain energy use criteria (e.g., KWH/ft^2). Although this ensures that new construction is held to the highest standards, it does not affect the existing building stock, which can make up a significant portion of the structures in existing communities.

The City of Berkeley, California, has pioneered an approach that reaches beyond new construction to gradually increase the average energy efficiency of the existing building stock, and relies on existing "off the shelf" technologies.[8] For over twenty years, Berkeley has required existing residential structures to meet certain energy-efficiency standards before being sold or undergoing renovations that will cost over $50,000. The Residential Energy Conservation Ordinance (RECO) requires energy-efficiency measures such as:

- increasing ceiling insulation to R-30;

- insulating hot water pipes and water heaters;

- installing low-flow showerheads, faucets, and toilets;

- using compact fluorescent lighting in common areas (in multi-unit residential buildings);

- applying weather-stripping to exterior doorways;

- sealing and insulating furnace ducts joints; and

- ensuring that fireplaces have dampers, doors, or other closures installed.

While many of the required measures are relatively low in cost, the law does cap an owner's financial obligations. Under RECO, owners are not required to spend more than 0.75% of the sale price of the property to complete the upgrades, or 1% of the cost of the major renovations (with alternative financial caps for multi-family buildings).

This law worked so well on the residential sector that Berkeley enacted a Commercial Energy Conservation Ordinance (CECO) in 1993.[9] Similarly to RECO, CECO requires energy-efficiency measures be undertaken before a commercial property is sold or undergoes major renovations. To comply with CECO, a building owner must undergo a thirty-two-point energy audit and comply with the findings of that audit (or file for an exemption).

Berkeley implemented a comprehensive decision-making process before moving ahead with the Residential and Commercial Energy Conservation Ordinances. By bringing citizens and the business community into the decision making, the city created an emissions reduction measure that financially benefits property owners, tenants, residents, and the city.

Table 4. **Impact of Berkeley's Residential Energy Conservation Ordinance**

Over 20,000 residences (50% of Berkeley's housing stock) increased energy efficiency

Over 130 commercial buildings (10% of city's total) increased energy efficiency

Residential natural gas use has declined 18% per capita

SOURCE: Excepted from International Council for Local Environmental Initiatives (ICLEI), *Best Practices for Climate Protection: A Local Government Guide* (Oakland, Calif.: ICLEI, 2000).

Santa Barbara, California, also has taken an innovative approach to encouraging energy efficiency in new construction. It can be a long and cumbersome process for contractors and developers to secure the permits necessary to start a construction project, and Santa Barbara found it could use its permitting process to encourage energy-efficient construction. Rather than mandating energy efficiency, the city implemented a voluntary, incentive-based program based on expedited permitting. Projects exceeding California's energy code, Title 24 (one of the strictest energy codes in the nation), by 15% are considered first by the permitting office. As Title 24 is updated continually, projects applying for the city's priority-review program also must become more efficient over time.

By expediting the permitting process, projects can move much faster from planning to construction phases. For developers and contractors, whose profit margin depends on speed and volume, significant value may be gained from expedited permitting. The community benefits by recognizing lower utility bills and corresponding decreases in greenhouse gas emissions. Since the program was instituted, the city has noted an increase in the number of projects submitted for review that are significantly more energy efficient than Title 24 requirements.[10]

Additional Resources

Building Code Assistance Project
202-530 2200
www.solstice.crest.org/efficiency/bcap

City of Berkeley's Residential and Commercial Energy Conservation Ordinances
www.ci.berkeley.ca.us/sustainable/residents/ResSidebar/RECO.html
www.ci.berkeley.ca.us/sustainable/buildings/ceco.html

Energy Efficiency and Renewable Energy Network
U.S. Department of Energy
1-800-363-3732
www.eren.doe.gov

United States Green Building Council
www.usgbc.org

Utilizing Landfill and Sewer Gas
as a Source of Green Power

Often, landfills are a community's single-largest producer of greenhouse gases, releasing methane as organic matter decays. As methane is a more potent greenhouse gas than CO_2 capturing the methane released from landfills, even if just to flare it (which converts the methane back to CO_2) provides local governments with an opportunity to significantly reduce their emission levels. If the captured landfill gas is then used to produce electricity, it reduces the need to generate energy from fossil fuels. This provides an additional reduction in greenhouse gas emissions and creates a financial resource from an often-overlooked source.

Austin Energy is a municipal utility that purchases electricity produced from landfill methane generated in the city and around the state. In 1995, the City of Austin established a methane-to-electricity facility at the Sunset Farms landfill in collaboration with a private contractor, who agreed to supply electricity to the city at a fixed rate for ten years.[11] This not only ensures consistent utility billing for consumers, but also provides customers with protection against rising energy costs. The success of the Sunset Farms facility has encouraged Austin Energy to explore opening similar facilities in other cities. In 2004, the city purchased 12 megawatts of electricity (at a rate of $40/MW) from Sunset Farms and a second plant outside of San Antonio.[12] Mark Kapner, a senior strategy planner for Austin Energy believes that "every landfill above a certain size is eventually going to be tapped. It's almost a no-brainer."[13]

Table 5. Outcomes of Landfill Gas Collection
at Sunset Farms

Sunset Farms generates approximately 3 MW of electrical power—enough to supply 2,000 homes

Prevents 2 million cubic feet of methane gas from being released from the landfill daily

Austin Energy's full Green Choice program offers 665 million kWh of green power to consumers

SOURCE: Excepted from International Council for Local Environmental Initiatives, *Best Practices for Climate Protection: A Local Government Guide.*

Similarly, a jurisdiction can capture the methane released from the breakdown of bio-solids in a sewage treatment plant. The methane captured can be used directly on-site in place of natural gas or can be burned to produce electricity. Portland, Oregon, has gone one step further and is utilizing an advanced technological solution to curb the greenhouse gases from the Columbia Boulevard treatment plant. Instead of burning the biogas the facility produces, it is processed (to remove impurities) and the hydrogen is extracted and used to operate a 170 KW fuel cell. This power station is able to reduce just less than 736 tons of greenhouse gas emissions annually (when compared to using a fossil fuel energy source) and it prevents the escape of methane into the atmosphere.[14]

Landfill and sewer gases are considered green energy sources because they come from biological sources (as opposed to fossil). Local governments can use these to offset internal energy use and emissions. If enough electricity is produced, it also can be made available to the public. Green energy can be sold at a premium to consumers seeking to lower their environmental impact. Austin Energy's "Green Choice" program gives consumers the option of purchasing green power at fixed rates. When the program was launched in 2000 there was a 5% premium for participating, but consumers were shielded from future price fluctuations for ten years.[15] The Green Choice program is one of the largest offerings of green power to public consumers in the country. In 2006, 665 million kilowatt-hours of clean power were sold to consumers from landfill gas, wind power purchases, and local solar power installations. The Green Choice program has been so successful that consumer demand for green power has outstripped supply and, as of December 2006, Austin Energy was not accepting new subscribers and was actively seeking new sources of green power to offer to their customers.[16]

Additional Resources

Landfill Methane Outreach Program
U.S. Environmental Protection Agency
1-888-782-7937
www.epa.gov/lmop

Encouraging Residential Recycling

Although landfill methane can be captured and used as an energy source, a greater emissions reduction can be achieved by preventing waste from entering the landfill in the first place. The average American is responsible for generating approximately 4.5 pounds of municipal solid waste daily.[17] One of the most effective means for reducing residential waste is to offer curbside recycling. As discussed in Chapter Three, recycling prevents greenhouse gases from being emitted through both direct and indirect manners—by preventing the direct release of carbon dioxide and methane (from decomposition), and by avoiding the energy use needed to produce new goods from raw materials.

Although almost 60% of the municipal solid waste produced in the United States is made up of recyclable materials, the national recycling rate hovers at approximately 24%. Similarly, compostable materials such as yard trimmings and food scraps account for another 25% of the municipal waste stream but are only recovered at a rate of 8.4% (for a combined recycling/composting rate of 32% nationally). Despite the potential that exists to reduce emissions and minimize the space needed for landfills, only half the U.S. population has access to curbside recycling programs.[18] These figures speak to an opportunity for local governments to make a positive impact.

Offering recycling as an option for a community is an obvious first step. Similarly, outreach and educational programs play large roles in local communities' participation rates in waste-reduction programs. Local governments also have an opportunity to take a more forceful (and potentially more effective) role in diverting waste streams from traditional landfills.

Seattle, Washington, a city that already exceeded the national recycling rates, aimed to decrease the amount of trash being sent to landfills even further through legislative action.[19] Although only about 25% of Seattle's garbage consisted of materials that could be recycled or composted (e.g., paper products, cans, bottles, yard debris), the city sought to target those materials and reverse a recent decline in recycling rates.[20] On January 1, 2005, a new city ordinance went into effect that banned the disposal of recyclable materials in trash bins collected by the city government. This ban extended to paper, cardboard, glass or plastic bottles and jars, aluminum or tin cans, and yard waste.[21]

To give the public ample notice of the new ordinance, the city launched an intensive education campaign and did not begin enforcement until 2006. Seattle Public Utilities began outreach to citizens with direct mailings and the establishment of an informational hotline in 2004. Starting in 2005, educational notice tags were placed on garbage cans containing significant amounts of banned waste types. "Significant amounts" was defined as "more than 10% by volume of container, dumpster or self-haul vehicles load based on visual inspection by an SPU inspector, contractor or transfer station worker."[22]

Seattle has taken a "carrot and stick" approach to increasing recycling rates, improving service as well as enforcing the ban. For example:

- free curbside recycling is available;

- yard waste and other compostables (vegetable food waste and compostable paper) is collected biweekly;

- food service businesses are provided a range of collection containers and pick-up frequencies at a reduced cost; and

- 300 new public recycling containers were installed in business districts throughout the city.[23]

Since the ban has gone into effect, the city's garbage collection contractors will not collect trash cans if they find large amounts of recyclable materials. Instead, they leave a tag on the can instructing the owner to separate the recyclables before the following week's collection. To enforce the ordinance on commercial establishments and multi-family residences, staff from the public utility will mail two warning notices to the holder of the garbage pickup account before adding a $50 fee to their garbage bill.

The ban on recyclables in municipal garbage collection not only reduces greenhouse gas emissions, but also increases the revenue generated from the sale of recyclables and saves residents money. Since there is a charge for waste collection, but not for recycling collection, the ban saves the public an estimated $2 million annually.

While some groups expressed concern regarding the ban on recyclables, the general public response has been overwhelmingly positive, and city outreach efforts have helped alleviate worries.[24] Between January and November of 2006, it is estimated that there has been a 99.98% compliance rate with the new law and that, due to a strong local market for recycled materials, the city itself saves $4.4 million annually through recycling waste as opposed to landfilling.[25]

Reducing Commercial Waste

Most municipal recycling programs generally focus on the residential sector because commercial waste collection is often provided by private contractors rather than municipal operations. However, this approach overlooks a significant portion of the waste generated within a community. Commercial, institutional, and industrial waste comprises 35 to 45% of the municipal waste stream, and these sectors produce the goods and packaging that eventually find their way into the residential waste stream.[26]

The State of Oregon operates a Commercial Waste Reduction Clearinghouse to deliver information on waste prevention, reducing consumption, environmentally preferable purchasing, and ultimately landfilling less waste. Resources include:

- information for businesses on setting up new waste reduction programs;

- lists of waste-reduction strategies;

- educational and promotional materials for encouraging involvement;

- resources, contacts, and funding opportunities;

- an overview of commercial laws and regulations; and

- case studies and success stories that others can draw from.[27]

This service makes it easy for local businesses to take steps to reduce the waste they generate without having to start from scratch in each case. Local governments around the country have developed similar outreach programs on a community or regional basis.

The Solid Waste Management Plan adopted by Bergen County, New Jersey, requires commercial establishments to recycle cardboard, paper, glass containers, cans, iron scrap, appliances, and construction and demolition debris. They must also track and report the amounts of materials recovered. This program has helped the county achieve a 63% diversion rate in industrial and commercial waste. To improve compliance, the county has created in-depth training and educational resources. The success of this program is assisted by the relatively high costs of waste disposal and a strong local market for recovered materials.[28]

Promoting Reuse as a Waste-Reduction Alternative

Many cities have a difficult time eliminating materials from the waste stream that do not fall into the easily sortable and recyclable classes of paper, glass, plastics, and metal. Consumer goods, many of which still have useful life left or contain toxic chemicals, often are put out on the curb, creating collection and disposal issues for waste collection departments.

To address these issues, the City of Saint Paul, Minnesota, launched the Twin Cities Free Market, Minnesota's first residential waste exchange. This program consists of an on-line database and web page that residents can use to list items they no longer want and to look for items they need. Interested parties can browse the listings and then contact the donor to arrange a pickup time. The Free Market is operated by Eureka Recycling, which also offers this service to other communities looking for ways to reduce waste. The program has prevented 2,000 tons of goods from reaching area landfills, extending the useful life of many goods and reducing greenhouse gas emissions in the process.[29]

In recent years, the popularity of such programs has been growing. Craig's List is available in many cities worldwide and has no charge for posting items for sale or to be given away. Excess Access's on-line system for matching business and household items with nearby nonprofits and recyclers has diverted 10,814 tons of materials from landfills (as of August 2006).[30] The Freecycle Network incorporates similar concepts to connect individuals within communities around the world.

In a similar vein, many jurisdictions have developed e-waste collection programs as an integral part of reaching their waste-diversion goals. These programs collect old computers, televisions, and other electronic items for free or at a minimal cost. Computers can be refurbished for nonprofit com-

Table 6. **Results from Sample E-Waste Collection Programs**

Jurisdiction	Average Collected	Average CO_2 Reductions
Fort Collins, Colorado	103 tons/year	308 tons/year
Madison, Wisconsin	56 tons/year	168 tons/year
St. Paul, Minnesota	55 tons/year	165 tons/year

SOURCE: Collection total were reported by city staff in 2004. Averages and emissions reductions were calculated by Ryan Bell using the methodology for computing emissions savings for recycling electronic waste outlined in the EPA's Waste Reduction Model (WARM).

munity groups, and the rest is recycled. These programs keep bulky waste out of landfills, allow for the proper collection of toxic materials, and realize the energy-reduction benefits associated with recycling the materials that make up the e-waste.

Additional Resources

City of Fort Collins Computer Recycling Website
www.ci.fort-collins.co.us/recycling/computers-recycle.php

Craig's List
www.craigslist.org

Excess Access
www.excessaccess.com/

Freecycle
www.freecycle.org/

Twin Cities Free Market
www.twincitiesfreemarket.org/

Conserving Water

Providing quality water to its citizens is a challenge many municipalities face. In most cases, surface and groundwater must be filtered and treated to improve quality before it can be consumed safely. The water then must be pumped to the site of use in homes or businesses, and wastewater must be pumped to treatment plants before being released back into the local hydrologic system. This is an expensive and energy-intensive process that accounts for approximately 35 percent of a local government's energy usage.[31] According to the American Council for an Energy Efficient Economy, water treatment and distribution nationally requires 50,000 GWH of electricity annually (1.4% of overall national consumption) at a cost of $4 billion.[32] While many people know that lower flow rates will save water, a commendable act in itself, few realize that these fixtures also reduce greenhouse gas emissions by decreasing the amount of water that must be treated and pumped throughout the jurisdiction.

The national Energy Policy Act of 1992 set minimum standards for water fixtures. Today's toilets cannot use more than 1.6 gallons of water per flush. This is a significant improvement over the pre-1992 standards, which allowed 4 to 6 gallons per flush. However, many of those older, water-intensive toilets are still in use throughout the country. Low-flow toilet rebates provide an incentive to property owners to replace their inefficient toilets with more efficient models. This saves homeowners money on their water bills, and saves the city money by reducing the energy costs of pumping and treating wastewater. Low-flow fixtures have a far-reaching effect, decreasing overall water consumption, the amount of wastewater treated and discharged back into the environment, and the emissions generated from the energy used in the supplying and treatment of the water.

Dual-flush toilets are another option for reducing water use. These fixtures give the user an option of 0.8 gallons per flush for liquid waste or the standard 1.6 gallons per flush for solid waste. This cuts water use in half every time the 0.8-gallon option is used. Commercial applications can take low-flow fixtures a step further with waterless urinals. As the name implies, these fixtures require no water to operate. A unique gel sealer allows urine to go through but no odors to come back up, creating a low-maintenance, completely waterless fixture. Some cities, such as Tybee Island, Georgia, are requiring waterless urinals in all public building restrooms.

Savannah, Georgia, has had a successful toilet exchange and low-flow fixture program in effect since 1998. Through this program, homeowners (or renters) come to a central location, fill out information about various fixtures in their houses, and receive low-flow toilets, faucet aerators, and showerheads to replace their existing fixtures. Homeowners are then required to bring old toilets to a drop-off site to avoid having any fees billed to them (and to prevent resale of the toilets). The first toilet swap was held on October 17, 1998, when four hundred ultra low flow toilets were distributed to residents who had pre-registered. Walk-ups were also accepted. Since that time, 1,270 single residential homes and 458 public housing units have been retrofitted with ultra low flow toilets, saving an estimated 7.1 million gallons of water per year. The average water savings per household from installing an ultra low flow toilet is 34 gallons per day and residences that have received ultra low flow toilets have shown a decrease of 16 to 24% in their water bills.[33]

Additional Resources

City of Savannah Water Conservation Department
www.savannahga.gov
912-651-4241

Encouraging Fuel-Efficient Vehicles

With gas prices routinely reaching historic high levels (breaking the $3.00 per gallon mark in September 2005 and multiple times during the summer of 2006), the demand for high-efficiency vehicles is increasing.[34] Filling this need are hybrid vehicles, which are becoming increasingly popular throughout the country. Throughout 2006, for example, there were month-long waiting lists to purchase the hybrid Toyota Prius. Hybrid cars use a smaller gasoline-powered engine in combination with an electric motor, allowing the car to run on electricity at low speeds and only requiring gas at higher speeds. The electric motor is operated off a battery pack, which is recharged when the gasoline engine produces more power than is needed to operate the vehicle and by storing the energy created when the vehicle brakes.

Hybrid vehicles achieve approximately 50 to 60 miles per gallon (mpg), comparing favorably with the 20 to 40 mpg of equivalently sized vehicles. Every gallon of gasoline burned results in over twenty pounds of carbon dioxide emissions. Therefore, switching from a traditionally powered vehicle to a hybrid can lead to a decrease of 3 tons of greenhouse gas emissions per vehicle, per year.[35] Although this may seem minimal on a per vehicle basis, if only a small percentage of the 136 million vehicles registered in the United States in 2004 were converted to hybrids, a significant reduction in greenhouse gas emissions could be achieved.[36]

Some local governments are following the lead of San Diego and Los Angeles, California, and rewarding citizens by offering free parking at city meters to drivers of hybrid vehicles or those that run on alternative fuels. Some local governments are also lobbying their state departments of transportation to permit hybrids to drive in the carpool/high-occupancy vehicle lane, a practice recently adopted in California. Together, these actions reward hybrid owners by allowing them to travel faster, at a more consistent speed (further increasing fuel economy), and to receive free parking at their destination. In order to provide a financial incentive, other municipalities are beginning to offer lower vehicle registration fees and taxes.

Reducing Single-Occupancy Vehicle Commuting with Parking Cash-Outs

The transportation sector is one of the largest sources of greenhouse gas emissions from cities and counties, and a large portion of these emissions are generated by single-occupancy vehicles used for commuting to work. Therefore, incentives to reduce emissions by getting commuters out of their vehicles and into alternative transportation mechanisms have become some of the more effective actions that can be taken at the local level.

The City of Santa Monica, California, enacted Ordinance 1604 to reduce traffic congestion and improve air quality. Under the ordinance, employers with ten or more employees must help reduce employee commute trips. Employers with ten to forty-nine employees are required to provide information to employees regarding air-quality issues, single-occupancy vehicle commuting, and ridesharing, and those with fifty or more employees must provide incentives for ridesharing and other strategies, as well as track progress in reducing solo commutes. Those employers with more than fifty employees that subsidize parking for their employees must prepare a parking cash-out plan that reimburses employees who choose not to drive to work an amount equivalent to the cost of providing them a parking space.[37]

A parking cash-out is a plan to offer compensation to employees who willingly give up their subsidized parking space. This financial incentive is a strong motivator for employees to find ways to carpool, rideshare, or take mass transit. Cash-outs range from $600 to $900 per participating employee annually. The program has seen a great deal of success, with approximately

Table 7. **Results of Santa Monica's Parking Cash-Out Program**

---

26 out of Santa Monica's 105 employers with 50 or more employees had implemented cash-out programs by February 1999

Vehicle miles traveled reduction of 544,000 miles per year

CO_2 emissions reduction of 196 tons per year

SOURCE: Excepted from International Council for Local Environmental Initiatives, *Best Practices for Climate Protection: A Local Government Guide.*

25% participation by employees in the Santa Monica cash-out program. As traffic congestion continues to grow and the price of parking continues to increase in cities across the country, parking cash-outs are likely to become a more popular strategy for reducing congestion, greenhouse gas emissions, and commuting times.

Additional Resources

City of Santa Monica Parking Cash-Out Program
www.santa-monica.org/planning/transportation/
abouttransmanagementtmo.html

Promoting Bicycling

Seattle, Washington, consistently ranked among the number-one cities for bikers in the country, has four to eight thousand bicycle commuters daily. Seattle's bicycle infrastructure consists of 28 miles of bike/pedestrian trails, 22 miles of on-street bike lanes, and about 90 miles of signed routes.[38]

Citizens who use the bicycle infrastructure advocate for improvements and have a Bicycle Plan Citizens' Advisory Board. In 1998, the city developed a Transportation Strategic Plan, which strongly advocated biking due to the numerous environmental and traffic-related benefits. The plan included a variety of initiatives, such as connecting urban trails to regional trails, providing adequate bike parking in buildings, repairing bike lanes, and working to ensure funding mechanisms for programs. The plan has realized significant results. Survey results of one prominent project indicated that the addition of bicycle lanes to major streets into downtown resulted in the reduction of 14,500 commuter vehicle miles and eliminated the release of 7 tons of carbon dioxide and 200 pounds of carbon monoxide.[39]

In 2006, Seattle's Department of Transportation began working on updates to the Bicycle Master Plan to further increase bicycle safety, increase bicycle use, and guide future improvements to Seattle's bicycling network. The new plan includes adding more bike lanes, signage, and maintenance plans. Public comments on the plan are underway and continuing into 2007.

Additional Resources

City of Seattle Bicycle Master Plan
www.cityofseattle.net/transportation/bikemaster.htm

Curbing the Urban Heat-Island Effect

Urban areas tend to be 10°F warmer, on average, than the surrounding countryside. These higher temperatures influence human health and comfort, energy used for air conditioning, and local air quality. The warmer air in these "heat islands" promotes the formation of ground-level ozone and smog and prolongs and intensifies heat waves in cities, increasing the rate of heat exhaustion and strokes. The "urban heat-island effect" is the result of:

- dark-colored building materials (e.g., asphalt shingles and roads), which cause an increased absorption of sunlight;

- increased heat storage due to the thermal properties of building materials;

- heat sources such as electricity use and fuel combustion; and

- less evaporation and lower wind speeds to disperse heat.

Cool Roofing. Conventional dark-colored, flat roofs absorb 80 to 90% of incoming solar radiation as heat, which warms the surrounding air and is transferred into the building, increasing demand for air conditioning. On hot days, conventional roofing materials can be 50 to 60° F hotter than cool roofing alternatives. Reflective roofing materials can lower these temperatures dramatically. For example, according to the U.S. Climate Change Technologies Program, energy savings in buildings with highly reflective roofs can be as high as 15% over the course of a summer, with savings as high as 32% during peak energy demand periods.[40]

Cool roofing strategies are in place around the country. When the City of Tucson, Arizona, applied a white roof coating to the city's service center, energy usage decreased by almost 50%—approximately 117,200 kWh. The city also saved $4,000 per year in energy costs and reduced their annual greenhouse gas emissions by 88 tons.[41] In January 2003, Chicago, Illinois, amended its energy code to require that low-sloped roofs on most new buildings reflect at least 25% of incoming solar radiation.[42]

Similarly, a vegetated "green roof" can minimize roof temperatures while advancing a community's aesthetic and ecological goals. One of the most famous U.S. examples of this technique is Chicago's City Hall, which is covered by 32,000 square feet of native grasses, shrubs, and trees. On a typical summer afternoon, the green roof is 50° F cooler than

an adjoining asphalt roof. The rooftop garden saves 9,272 kWh of electricity and 7,372 therms of natural gas annually, which translates into an
annual energy savings of $3,600. These energy savings also lead to an
emissions reduction of 46 tons of greenhouse gases, 19 pounds NO_x, 20
pounds SO_x, 33 pounds CO, 7 pounds VOCs, and 5 pounds PM_{10}. The
roof is also expected to retain 75% of the rainfall that occurs during a
one-inch storm event, preventing 1,250 cubic feet of runoff from reaching the sewer system.[43]

Lightening Streets and Cooling Parking Lots. Streets and parking
lots constructed using black asphalt cover a large portion of the surface of
an urban area. These dark-colored pavements can become 40° F hotter
than the surrounding air.[44] Constructing, replacing, or covering roads and
parking lots with reflective or cool paving materials, such as Portland and
flyash cement concrete, porous concrete, chip-seals, turf-block or porous
pavers, and light-colored asphalt emulsion seal-coats, have been demonstrated to be an effective way to lower pavement surface and ambient air
temperatures.

Greening the Community. Mature trees and other vegetation can keep
communities 7° F cooler than similar treeless areas by releasing water into
the air and by shading heat-absorbing surfaces.[45] Trees growing on the
west, northwest, and eastern sides of buildings can reduce cooling costs in
that structure significantly. Energy-saving simulations for Sacramento and
Phoenix found that three mature trees can cut home air conditioning
demand by 25 to 40%.[46] Preserving and increasing a community's overall
tree cover also can decrease electric bills. A study of twenty Miami and
Ft. Lauderdale neighborhoods determined that neighborhoods with more
than 20% tree-canopy coverage had summer electric bills 8 to 12% lower
than neighborhoods with less coverage.[47] Planting shade trees in hot spots
such as parking lots can have additional air quality benefits. An increase in
parking lot tree cover from 8 to 50% reduces the evaporation of hydrocar-
bons from fuel tanks and lowers the emissions of NO_x from vehicle start-
ups.[48] Chicago's landscaping ordinance encourages landscaping with trees
and shrubs in parkways and to screen or shade parking lots, loading docks,
and other vehicular use areas. Similarly, several central California cities,
including Sacramento and Davis, require that parking lots achieve 50%
shade cover within fifteen years of being constructed.[49]

Additional Resources

City of Chicago Environmental Action Agenda
www.cityofchicago.org/Environment/

EPA Heat Island Reduction Initiative
www.epa.gov/heatisland/

ICLEI's Urban Heat Island Reduction Program
www.hotcities.org

Lawrence Berkeley National Laboratory's Urban Heat Island Group
http://eande.lbl.gov/HeatIsland/

Chapter Six

Governments Gone Green

This is a matter of life and death for future generations
and to this planet.[1]

Mayor John D. Medinger, La Crosse, Wisconsin

Why should the community work to reduce emissions if local leaders are
doing nothing to limit their own administration's greenhouse gas contributions?
Although community-based efforts are long-lasting, far-reaching, and can generate popular support, local governments can and should be leaders by altering
their operating procedures to reduce greenhouse gases. This sets an example
of the government's commitment, and may generate additional support for
broader, community-based initiatives. Additionally, it is often easier to change
municipal policy internally before rolling out community-wide initiatives.

Green Building Policies for Municipal Buildings

As mentioned in Chapter Five, buildings consume almost one-third of
energy in the United States, and this energy use is responsible for 35% of
the country's CO_2 emissions. The U.S. Green Building Council is a coalition
of corporations, builders, universities, government agencies, and nonprofit
organizations working together to promote buildings that are environmentally responsible, profitable, and healthy places to live and work. Since its
founding in 1993, the council has grown to include more than 7,500 members and a network of seventy-five regional chapters that work to incorporate sustainability into the built environment.[2] The U.S. Green Building
Council is perhaps best known for their portfolio of products falling under
the Leadership in Energy Efficient Development (LEED)® program.

The LEED Green Building Rating System® is a feature-oriented rating
system that gives credits for satisfying specified green building criteria.
Based on the total credits earned, buildings receive Certified, Silver, Gold,
or Platinum levels of green building certification. The LEED system focuses
on five primary areas—site issues (including transportation, stormwater, and

the heat-island effect), water use (both in the building and for irrigation), energy efficiency and the source of electricity generation, materials used to construct the building, and indoor environmental quality (air quality, thermal comfort, daylighting).

To date, over fifty cities and counties have passed ordinances that all new municipal construction has to meet the LEED standards.[3] Municipal buildings present excellent options for reducing both energy and water demand, as well as creating higher performance work areas and healthier working environments through increased natural lighting and better indoor air quality. By committing to incorporating green building standards into new buildings, local governments realize a payback over the life of the building from much lower operating expenses, as well as contributing to the increased health and well-being of their employees.

LEED buildings reduce greenhouse gas emissions in numerous ways. According to the Green Building Council, the average LEED building is 30% more energy efficient than a building built to code.[4] The LEED standards also encourage the use of renewable energy or green power and alternative forms of transportation by awarding credits for offering bicycle storage and showers, preferred carpool and hybrid vehicle parking spaces, and being in close proximity to mass transit. Reduced water use (averaging 40% in LEED buildings) decreases energy use by municipal water and sewage treatment plants.[5] LEED also rewards the use of local and recycled materials, which reduces transportation emissions (and costs) and helps to stimulate product demand. Finally, LEED certification requires that building occupants are offered recycling opportunities, thereby reducing landfill methane emissions.

While LEED is one popular standard, many local governments have created their own set of green building criteria for use in municipal facilities. Beginning in 1995, Tucson and Pima County, Arizona, jointly established a minimum energy-efficiency standard for building construction based on the Model Energy Code of 1995 (now called the International Energy Conservation Code), which set national minimum energy-efficiency standards for building insulation, windows, lights, and other components.[6] In 1998, the City of Tucson increased this standard, requiring newly constructed and significantly renovated municipal buildings to meet the "Sustainable Energy Standard," which has energy-efficiency requirements 50% higher than the Model Energy Code. The Sustainable Energy Standard was again updated in 2003 to reflect changes that made the energy codes, on which it is based, more stringent (based on ASHRAE 90.1-2001).

Table 8. **Annual Results from Tucson's Sustainable Energy Standard**

$73,000 per year saved through avoided utility costs

784 tons of CO_2 reduced annually

Annual savings will grow as more renovation and construction are completed

SOURCE: From International Council for Local Environmental Initiatives, *Best Practices for Climate Protection: A Local Government Guide.*

Originally developed for a solar neighborhood in Tucson called Civano in the early 1990s, the Sustainable Energy Standard is mandatory for municipal construction and voluntary, but encouraged, for private developments. The Sustainable Energy Standard suggests various conservation measures, but allows design professionals to choose the strategies that work best for each specific project. Energy modeling, a key component used to demonstrate the various energy-efficiency techniques, also illustrates the energy cost savings associated with each measure. The city then monitors energy efficiency throughout the contracting, inspection, and testing phases. This process ensures that all involved, from design to construction, understand the importance of energy efficiency, and ensures that the savings are realized.[7]

Additional Resources

United States Green Building Council
www.usgbc.org

Comprehensive Upgrade of City Buildings

When new construction is unnecessary, retrofitting existing municipal buildings not only reduces greenhouse gas emissions, but also provides an attractive financial incentive. Older buildings often are equipped with inefficient lighting, heating, and cooling mechanisms, poor ventilation, inadequate insulation and glass, and a host of other operational issues. In older buildings, these upgrades rapidly pay for themselves. For example, old-style T-12 fluorescent lights and magnetic ballasts typically are 40 to 60% less efficient than the newer, readily available, T-8 electronically ballasted light fixtures. The energy savings can pay for the upgrade in only a few years, and the newer fixtures generally produce a better quality of light as well.

The U.S. Green Building Council also has developed LEED standards for existing buildings. The standards provide guidance for improving the energy efficiency of building operations and other systems without making major changes to the interior and exterior of the building. Specifically, the standards address exterior building site-maintenance programs, water and energy use, environmentally preferred products for cleaning and alterations, waste management, and ongoing indoor environmental quality.

In recent years, cities around the country have made major strides in improving the energy efficiency of existing city office buildings, schools, and police and fire stations. Energy-efficiency measures often are the first step local governments take when upgrading their facilities because they offer direct financial paybacks through reduced energy consumption while also reducing greenhouse gas emissions.

The City of Toledo, Ohio, took steps to reduce their energy consumption by retrofitting twenty city-owned buildings and facilities.[8] The older buildings required numerous energy-efficiency measures that were undertaken easily and offered a quick return on investment. Toledo retrofitted a diverse building stock—from municipal courthouses to police and fire stations to parking garages. Measures included upgrading to efficient lighting with motion sensors and installing new boilers and chillers, which eliminated the use of ozone-depleting coolants and replaced old window-mounted air conditioning units. Centralized building automation systems were also installed

Table 9. **Toledo's Building Retrofit Program—Results in the First Year**

20 city buildings retrofitted
Cut electricity use by 5,823,000 kWh
Cut natural gas use by 111,892 hundred cubic feet
Reduced 5,250 tons CO_2
Saved $710,208 in the first year

SOURCE: Excerpted from International Council for Local Environmental Initiatives, *Best Practices for Climate Protection: A Local Government Guide.*

as part of the upgrades to assist maintenance staff in diagnosing comfort and maintenance issues in a timely manner.

The city contracted with an energy service company, which developed a program to ensure that the cost of the improvements would be covered by the future energy savings, ensuring that there was ultimately no net expense to the municipality. Energy service companies provide an effective means for municipalities to identify and finance comprehensive building upgrades. These companies often perform an initial audit of the buildings for free. Through examinations of past utility usage and cost data as well as site visits, these companies suggest a range of mechanical, electrical, and plumbing upgrades that not only improve energy performance but also significantly reduce operating costs. Local governments can then choose which strategies to pursue to meet the needs for each individual building.

These packages usually are tied to the expected financial payback. The building owner/operator looks at the cost savings, analyzes the payback, and identifies which improvements will produce the largest return on investment. Often contracts with energy services companies allow repayment for the improvements to be made over time from the realized saving. Some companies even go as far as linking their payment to the cost savings that are achieved. If the savings are less than anticipated, the company receives a lower payment for its services.

Local governments that own buildings that have been in existence for several years can use these relationships as springboards for energy-efficiency measures. By outsourcing the first cost estimates and payback schedules for the various performance upgrades to industry experts, government officials also can eliminate the multiple staff hours required to do this analysis in-house.

Additional Resources

Energy Star Buildings
U.S. Environmental Protection Agency
1-888-782-7937
www.energystar.gov

"Steps to Successful Municipal Energy Management"
Climate Institute
202-547-0105
ncwilson@climate.org

United States Green Building Council
www.usgbc.org

Green Power for Municipal Operations

While local governments often find an acceptable return on investment through energy-efficiency measures, it is sometimes harder to justify direct investments in the installation of new renewable energy systems. Electric utilities increasingly are offering governments and consumers the opportunity to choose the source of their electricity, by aggregating demand and using market forces to reduce the cost of developing new generation sources to the individual consumer. In 1999, the City of Santa Monica took advantage of electricity deregulation to become one of the first cities in the country to purchase 100% renewable electricity for all of its facilities.[9]

Although city officials initially worried about costs and interest on the part of the utilities, they received over a dozen bids, and the cost premium to the city was only 5%. This made it possible for the city to switch from grid electricity (which contained only 11% renewable energy) to 100% geothermal electricity.[10] The city has been able to maintain this agreement and their favorable rates through California's energy crisis, renewing their contract in 2004 to continue to operate on 100% renewable, greenhouse gas–free electricity.[11]

In conjunction with the city's purchase, it also launched an educational campaign to promote green power and energy efficiency to residents, institutions, and energy customers served by the Los Angeles Department of Water and Power.[12] Surveys of residents and businesses found that almost three-quarters of respondents would be interested in purchasing renewable energy if the price differential were no more than 5%.[13] In leading by example, Santa Monica is able to talk to the community from a position of experience.

In 2004, Montgomery County, Virginia, led a group of local governments and local government agencies in a wind energy purchase that represents 5% of the buying group's total electricity needs. Under the two-year deal, the buying group collectively purchased 38 million kWh of wind energy annually, translating into a yearly eCO_2 reduction of approximately 21,000 tons.[14] The county demonstrated the benefits of renewable energy in meeting the requirements of the federal Clean Air Act by including the wind energy purchase as a control measure for ozone pollution in a "State Implementation Plan" for air quality improvement. The county offsets the added expense of the wind power purchase by instituting employee

Table 10. **Annual Results of Santa Monica's Green Power Purchase Program**

5 MW 100% renewable power purchased
Eliminates 13,672 tons CO_2
Cut 16.2 tons NO_x
Cut 14.6 tons SO_x
Cut 2,285 pounds PM_{10}

SOURCE: Excerpted from International Council for Local Environmental Initiatives, *Best Practices for Climate Protection: A Local Government Guide.*

energy-efficiency programs such as turning off lights, computers, and office equipment when not in use.

In municipalities where green power is not available from the utility, another option is to purchase renewable energy credits or "green tags." Simply stated, green tags are a decoupling of the environmental benefit from the actual power purchase. There are three basic premises behind green tags. First, renewable power still costs slightly more to generate than traditional fossil fuel–derived electricity. Second, it is not always possible to purchase green power in all markets. Finally, people are willing to pay extra for more environmentally benign power. Green tags separate the electricity purchased from the sale of the environmental benefits. Companies determine the amount of greenhouse gases saved by producing electricity from renewable sources and how much extra this power costs to produce, and then use this information to calculate the cost of the CO_2 savings. They can then sell the electricity at the market rate and sell the green tags to users who wish to purchase clean electricity but do not have that option. There are many brokers of green tags throughout the country. A starting point is with an independent third party—such as Green-E—that verifies that green tags being sold are actually produced and appropriately labeled, and ensures that brokers are not selling green tags to multiple individuals.

Additional Resources

The Green-E Renewable Energy Program
Center for Resource Solutions

415-561-2100

www.green-e.org

Biodiesel for Municipal Fleets

All across the United States, individuals, businesses, and governments are actively choosing a cleaner and more energy-independent future through the use of biodiesel, making it the fastest-growing source of alternative fuel in the country.[15] Thousands of U.S. fleets, businesses, and individuals operate using biodiesel. Fleet users of biodiesel include school districts, national parks, the military, local and regional bus systems, as well as an increasing number of local governments. In the last few years, a network of public and private biodiesel fueling stations has sprung up in all fifty states, increasing the accessibility of this alternative fuel for all users.[16]

Biodiesel is produced from recycled or virgin vegetable or animal oils and can be used in most diesel engines without undertaking costly modifications to the vehicle fleet or fueling infrastructure. Most biodiesel in the United States is made from virgin soybean oil, but recycled restaurant oil and grease also can be used in fuel production, thereby turning a waste product into a valuable resource. Moreover, as biodiesel can be produced from domestically available materials, its use contributes to national energy independence and economic growth.

Biodiesel can be used in its pure form (100% biodiesel, known as B100) or can be blended with traditional petroleum-based diesel in any proportion. The most commonly found blend of biodiesel is B20, a mixture of 20% pure biodiesel and 80% petroleum diesel. The performance, storage requirements, and maintenance of biodiesel are similar to, if not better than, that of conventional diesel. It is also far safer to transport, store, and use because it is not a hazardous material like petroleum-based diesel.

In addition to its waste reduction potential, biodiesel has significant air quality benefits. In 2000, it became the only alternative fuel in the country to have successfully completed the EPA-required Tier I and Tier II health effects testing under the Clean Air Act.[17] It is proven to be safe and biodegradable, and reduces the emission of serious air pollutants such as sulfur compounds, particulate matter, and carbon monoxide.[18] Because biodiesel is derived from biologic feedstock, as opposed to fossil fuel, it reduces carbon dioxide emissions by 78% (over its lifecycle) when compared to its fossil fuel equivalent.[19] Although studies have shown that the emissions of nitrogen compounds (NO_x) from biodiesel may be slightly higher than conventional diesel, additives are being developed to reduce those emissions.[20] The overall end-of-tailpipe emissions of air pollutants are decreased significantly by switching to biodiesel, and the total emissions benefits are even more dramatic when the full lifecycle of the fuel is considered.

The price of biodiesel depends on the location and size of the producer and distributor, as well as the percent blend of biodiesel with regular diesel. In the past few years, consumers expected to pay between 13 and 22 cents more per gallon of B20 than for petroleum diesel, but with recent increases in fuel prices, biodiesel has become more competitive.[21] In 2005 and 2006, the price of B20 has been competitive with fossil fuel–based diesel, though B100 still requires a 60 to 80% premium.[22] Additionally, because it uses existing infrastructure and vehicles without requiring expensive new equipment or retrofit, biodiesel may be a least-cost alternative for complying with state and federal air quality regulations requiring the use of alternatively fueled fleet vehicles.[23]

Las Vegas, Nevada, and the surrounding Clark County suffer from significant air pollution problems, exacerbated by their geography: a hot, sunny, desert environment surrounded by mountains that increase the rate of smog formation, which is then trapped in the valley. With the restaurants, casinos, and hotels in Las Vegas and Clark County producing 6 gallons of grease per resident per year (twice the national average), recycling that waste stream into an energy source was a natural way to tackle both waste disposal and air quality issues.[24]

Recognizing this fact, local agencies Haycock Petroleum and Biodiesel Industries were able to work with the Nevada Energy Office to take advantage of the state's interest in promoting biodiesel and a Department of Energy grant to open the first commercial refinery in the nation to operate primarily on a feedstock of recycled cooking oil.[25] In addition, this partnership opened the country's first public biodiesel fueling station in 2001. This resource made it possible for the City of Las Vegas, the Las Vegas Valley Water District, the Clark County School District, a local university, and the Air Force to integrate biodiesel into their fleets.

Together, these agencies create a significant base of alternative fuel users. The school district alone operates the largest biodiesel school bus fleet in the country, with 1,188 buses running on the fuel. In 2004, 2,288 government fleet vehicles consumed over 4,000,000 gallons of B20 annually, leading to the significant reduction in greenhouse gases and air pollutants included in Table 11.

In an effort to reduce NO_x emissions, the city has begun using a fuel

Table 11. **Emissions Reductions through the Use
of Biodiesel by Agencies in and
around Las Vegas**

8,798 tons eCO_2

28 tons CO

2 tons SO_x

7 tons VOCs

2 tons PM_{10}

SOURCE: Emissions reductions from fuel usage calculated by Ryan Bell using
the Clear Air Climate Protection Software Version 1.1.

additive in all engines running on biodiesel. This additive improves fuel
combustion, thereby potentially improving fuel efficiency by 3 to 5% and
reducing the air pollutants (soot, carbon, sulfur, and nitrogen compounds)
associated with incomplete fuel combustion. Although the additive adds 2
cents per gallon to the cost of the biodiesel, it has reduced NO_x emissions by
36%, or an estimated 111 tons over the course of a year, when compared
with conventional diesel.[26]

Overall, the biodiesel project has been successful, with high driver and
rider acceptance. Because about half of the biodiesel is produced by a local
biodiesel manufacturer from used vegetable oil, the city has eliminated the
disposal need for the waste, contributed to the local economy, and significantly reduced the emissions associated with the burning and transport of
petroleum diesel.

Additional Resources

Clean Cities Newsletter on Clark County School District Biodiesel Project
www.eere.energy.gov/afdc/apps/toolkit/pdfs/las_vegas_success.pdf

Las Vegas Regional Clean Cities Coalition
www.lasvegascleancities.org/

The National Biodiesel Board
http://www.biodiesel.org/

Green Municipal Fleets

Local governments can also help reduce emissions from the transportation sector by taking steps to clean up their own municipal fleets. In the early 1990s, the City and County of Denver, Colorado, faced soaring fuel costs and federal mandates to improve air quality in the region. In response, the mayor signed a "Green Fleets" executive order on Earth Day 1993. This order required government fleet managers to take steps to reduce CO_2 emissions and fuel costs by 1% per year over the following ten-year period.[27]

The program, which was updated in 2000 and 2003, calls for the city to minimize the environmental impact of its fleet and "enhance domestic security" through "maximize[ing] fuel efficiency and diversity."[28] According to the program objectives, this is to be accomplished through:

- increasing the fleet average fuel economy;

- minimizing the vehicle miles traveled by employees; and

- increasing the number of alternative fuel and hybrid fleet vehicles.[29]

Fuel consumption (and thus carbon emissions) is reduced by requiring fleet managers to purchase the most cost-effective and lowest-emission vehicle possible that still meets agency needs. In order to accomplish this goal, fuel-efficiency standards are included in procurement specifications, and old and underused vehicles are eliminated from the inventory. The Green Fleet program made large strides in 2001 when thirty-nine Toyota Prius hybrids were purchased. The uniquely shaped Toyota Priuses were visible signs of the city's commitment to fuel efficiency, and quickly became popular with city staff. By 2005, over eighty hybrid vehicles (including SUVs) were in the city fleet.[30]

A review committee, appointed by the mayor, monitors the performance of the Green Fleets program and tracks effectiveness by analyzing the corresponding fleet energy use and CO_2 emissions. Authorities estimate that Green Fleets activities take up a minimal amount of staff time, thereby helping to keep costs low.[31] Overall, the program has been a resounding success, and the city is continuing to set higher fleet standards—currently the program is experimenting with alternative fuels to further reduce its transportation emissions.

As Denver has demonstrated, greening the municipal fleet is becoming even easier with the escalation of fuel prices and the emergence of hybrid

Table 12. **1999 Results of Denver's Green Fleets Program**

Offset the city's fleet growth by 10 vehicles and
downsized 13 others

Saved $40,000 in operation and maintenance costs

Saved up to $100,000 in capital costs by not purchasing
some of the vehicles requested

Prevented the emission of 10 to 15 tons of CO_2

SOURCE: Excerpted from "Denver's Green Fleets Executive Orders,"
www.greenfleets.org/Denver.html.

vehicles that achieve 50 miles per gallon, almost twice the current federal
fuel-economy standards for passenger vehicles. Although there is a cost premium associated with purchasing hybrid vehicles, rising gasoline prices make
replacing standard fleet vehicles with efficient hybrids a more attractive way
to reduce greenhouse gas emissions. Federal tax incentives are also available
for purchasing hybrids that can further offset the cost of the new vehicles.

Fleet managers in Charlotte, North Carolina, determined that hybrids
would be less costly over their lifetimes than their current fleet vehicles. In
addition to saving gas, their higher resale value and reduced maintenance
costs help offset the premium paid on the initial purchase. In fact, Charlotte's
Fleet Environmental Analyst estimates that switching from a gas-only Ford
Taurus to a hybrid Toyota Prius or Honda Civic would save city taxpayers approximately $800 to $1,200 annually per vehicle.[32] With cost premiums typically under a few thousand dollars, this payback is less than five years. This
move also will result in an estimated reduction of 30 tons of carbon dioxide.[33]

Additional Resources

Green Fleet Resources
www.greenfleets.org/

The Green Guide to Cars and Trucks
American Council for an Energy- Efficient Economy (ACEEE)
202-429-0063
www.aceee.org

Innovative Paper Reduction Strategies

Paper use has a significant impact on both local and global environmental issues. Pulp and paper production is the fifth-largest industrial consumer of energy in the world, using as much power to produce a ton of product as the iron and steel industry. A majority of the country's timber harvest is used in the production of pulp for paper, and the pulp and paper industry has been cited in the release of dioxin and chlorine into the air and water. This all occurs to make a product that usually is used once and thrown away. In some countries, including the United States, paper accounts for nearly 40% of all municipal solid waste, making it the leading producer of methane emissions in this country. Therefore, efforts to reduce the amount of paper used and disposed of is an easy way for local governments to lower their greenhouse gas emissions while also minimizing pressure on forest resources and eliminating a significant source of air and water pollution.

Even cities that already have in-house recycling programs often find new ways to increase recycling rates. Overland Park, Kansas, took a hard look at waste disposal habits and noticed that city employees had much larger bins for disposable waste than they did for recyclables. Further analysis identified that only 10% of eligible recyclable materials were being recovered. The city decided to replace the older waste containers with their opposite, ones with a larger compartment for recyclables and a smaller compartment for waste. Following the switch, the amount of solid waste generated dropped by more than 70%, leading to a 30% decrease in waste disposal fees. These savings allowed the city to recover the cost of investing in new containers in under a year.[34]

Portland, Oregon, realizing the impact that excessive paper use has on natural resources, waste disposal, and the environment, has also worked to reduce its consumption in municipal offices.[35] The city adopted a Sustainable Paper Use Policy in June 2003. The policy calls for a number of measures aimed at reducing the amount and impact of paper use in city operations. The stated goals of this program include:

1. reducing paper consumption;
2. considering fiber source and type, paper processing methods, and recyclability in paper purchase decisions in addition to price, performance quality, and end-use application;
3. reusing and recycling paper products.[36]

To reduce paper consumption, the policy requires duplex printing on all printers, faxes, and copies, and eliminates the use of individual printers in favor of shared, centrally located printers. Paper type is addressed by requiring that all paper products meet the EPA recycled-content recommendations, giving preference to chlorine-free paper products, and ensuring that suppliers provide the source certificates for the various paper products. Finally, all bureaus are required to establish paper reduction strategies, and to track printing, copying, and writing paper consumption.

This stringent policy has already proven effective in reducing the city's paper use. In 2004, paper use is down 15%, cutting more than 13 million sheets (approximately 26,000 pounds) with an estimated reduction of 34 tons of eCO_2.[37]

Reducing paper use through modifying internal procurement, usage, and disposal policies is an effective means to lessen the environmental impact and associated greenhouse gas emissions of paper consumption. Other governments have turned to technology to reduce paper use while significantly improving the efficiency and cost-effectiveness of their operations. Examples include the transition to electronic filing systems and online or phone-based systems for providing public services. In addition to reducing paper use, these systems can expedite and streamline costly and time-consuming procedures, minimize vehicle travel to access services, and enhance government transparency, accuracy, and accessibility.

The traffic court in Miami-Dade County, Florida, is the fourth-largest in the world, processing the traffic citations issued by every law enforcement agency in the county.[38] Duties range from updating, organizing, and maintaining public records to collecting fines and maintaining calendars for county court judges and hearing officers. In 1995, the traffic division of the Miami-Dade Clerk of Courts, in cooperation with the Administrative Office of the Courts (AOC) and Miami-Dade County, introduced a "paperless" traffic court. The project was initiated to improve service, and in reaction to financial pressures to be able to "do more with less."

The SPIRIT (Simultaneous Paperless Image Retrieval Information Technology) system is a collection of programs, databases, and computers that digitizes paper-based documents, providing access to judges, attorneys, and court clerks throughout county facilities. Instead of paper files, SPIRIT courtrooms have imaging workstations to digitize information. The SPIRIT systems not only have decreased paper use but have increased worker productivity significantly. Before the new system was developed, clerks handled each piece of paper an average of thirty-seven times. SPIRIT currently handles up to thirty-six hundred cases a day, allowing the traffic division to process 32% more citations, despite a 167% increase in court cases, while lowering error rate from 15% to less than 1%.

This increase in efficiency also translated into financial savings—the division was able to transfer 15% of their workforce to other areas of the clerk's office while reducing overtime pay by almost $275,000 annually. Customer service has improved as well, as more time can be spent working with people instead of processing paperwork. Finally, the overall environmental impact of the court has been decreased, as the digital process allows the public to utilize other locations for traffic-related business (such as public service counters, attorney rooms, courtrooms), eliminating trips to the downtown location to review and process files. From saved money to reduced storage needs to support among the staff and the public, Miami-Dade's "paperless" traffic court demonstrates the diverse benefits of e-government.

In addition, Miami-Dade County developed and implemented an Interactive Voice Response telephone system known as DIAL (Direct Information Access Lines) to provide general information on traffic violations, court procedures, and court-approved programs. Citizens can pay for traffic and parking tickets, set court dates, or make child support inquiries over the phone or online. Since its inception, the system has handled as many as twelve thousand calls per month, reducing the need for processing considerable amounts of paperwork, which eventually would have resulted in additional paper waste. The DIAL system also significantly reduces the environmental impacts associated with transportation to and from the courts, eliminating an average of 951,064 vehicle miles and 578 tons of eCO_2 annually.[39]

Additional Resources

City of Portland Sustainable Paper Use Policy
www.portlandonline.com/shared/cfm/image.cfm?id=24521

The EPA's Comprehensive Procurement Guidelines (which contains recommendations regarding the recycled content of purchased paper products)
www.epa.gov/cpg/pdf/paper.pdf

Miami-Dade County Clerk of Courts DIAL program website
www.miami-dadeclerk.com/dadecoc/DIAL.asp

Miami-Dade County Clerk of Courts SPIRIT program website
www.miami-dadeclerk.com/dadecoc/SPIRIT.asp

Environmentally Preferable Procurement Programs

U.S. state and local governments spend $30 to $40 billion a year on energy-consuming products and equipment. Specifying energy-efficient products not only reduce ongoing utility costs but also cut significantly the associated greenhouse gas emissions. Fortunately, finding energy-efficient office products is as easy as requiring the Energy Star label on applicable equipment. Energy Star is a U.S. EPA and Department of Energy program to certify, label, and promote a diverse array of energy-efficient products and practices—ranging from light bulbs to photocopiers to energy management guidelines and financial calculators. The Energy Star label allows for easy recognition of the most energy-efficient and least-polluting products.

Beginning in 1997, the State of Massachusetts began requiring the purchase of Energy Star labeled office equipment. These rules not only require the purchase of Energy Star rated equipment, but also that vendors enable all power-saving features at the time of shipment to ensure that energy savings are realized. To increase understanding of the policy and help other governments and agencies adopt similar policies, the state has developed educational materials, provides trainings for technicians and employees, and assists in the tracking of savings to promote successes.[40]

Table 13. **Potential Impact of Purchasing Energy Star Office Equipment**

All office equipment and most appliances now purchased are energy efficient
Each Energy Star computer and monitor eliminates nearly 1 ton of CO_2 per year
Each Energy Star office product saves $15 to $25 per year in energy costs

SOURCE: Excerpted from International Council for Local Environmental Initiatives, *Best Practices for Climate Protection: A Local Government Guide.*

Local governments can go beyond energy-efficient purchasing and adopt "environmentally preferable purchasing" policies.[41] Environmentally preferable purchasing involves buying products or services that have a lesser or reduced adverse effect on human health and the environment when compared with competing products or services that serve the same purpose.[42] Purchasing agents must consider a host of different factors throughout the lifecycle of the products—from the impact of the raw material used, through production and transport and finally the ultimate disposal of the product. The EPA's "Final Guidance on Environmentally Preferable Purchasing" centers on five guiding principles:

1. including environmental considerations as part of the normal purchasing process;
2. emphasizing pollution prevention early in the purchasing process;
3. examining multiple environmental attributes throughout a product's lifecycle;
4. comparing environmental impacts when selecting products; and
5. making purchasing decisions based on accurate and meaningful information about environmental performance of products and services.[43]

Environmentally preferable purchasing provides a holistic look at the procurement process and can help governments become more environmentally and socially sustainable. From examining the recycled content of products to lessen the greenhouse gas emissions associated with raw material extraction (see "The Lifecycle of an Aluminum Can" in Chapter Three) to selecting materials that are manufactured locally to reduce transportation emissions, environmentally preferable purchasing is a far-reaching tool. Local environmentally preferable purchasing provisions are improving worker safety and health through the use of less-toxic products, reducing liability due to decreased risk of occupational hazards, reducing health costs associated with those hazards, and stimulating the market for environmentally friendly products. Such purchasing polices are visible demonstrations of the leadership positions that local governments are taking to reduce emissions and improve overall environmental quality.

Additional Resources

Energy Star Purchasing Toolkit
U. S. Environmental Protection Agency and Department of Energy

1-888-782-7937
www.energystar.gov

U.S. EPA Environmentally Preferable Purchasing Resources
www.epa.gov/epp/

Concrete with Recycled Content

More than half of the United States industrial sector's emissions of greenhouse gases and 5 to 8% of the world's CO_2 come from one industry, the manufacturing of cement.[44] Cement production requires the heating of limestone to extremely high temperatures to convert it to lime, the basic ingredient for cement production. During this process, large amounts of CO_2 are released from the limestone. This is further exacerbated by the cement industry's heavy reliance on coal to fuel this process. Each ton of conventional cement produced emits one ton of carbon dioxide. In addition to greenhouse gases, the cement industry also releases large quantities of nitrous oxide, sulfur compounds, and other air pollutants.

Fortunately, alternatives to the traditional cement blend in concrete can reduce greenhouse gas emissions substantially and minimize the use of virgin materials. Concrete containing recycled content (also known as blended cement) can utilize a variety of alterative binding materials to replace standard Portland cement. These alternatives reduce greenhouse gas emissions associated with the production of cement because they do not require the extensive "firing" process. They often include the byproducts of industrial processes that otherwise would be sent to a landfill. Such byproducts include fly ash (a residue from coal-fired power plants), slag (a byproduct of steel smelting), or rice hulls. Additionally, using recycled material can achieve a cost savings for a project. An analysis by the California Integrated Waste Management Board showed savings as high as 49% over the use of traditional cement materials.[45]

In December 2002, Berkeley, California, became the first U.S. city to adopt a blended-cement procurement policy. The policy requires that, wherever technically appropriate, procurement of cement will specify the use of blended cement in city buildings and other construction. Concrete with recycled content typically takes longer to cure than traditional concrete, so its use is not applicable in all situations. However, once cured, recycled content concrete is harder than traditional concrete and potentially more durable. Since fly ash and slag were readily available in Berkeley (and across the country), the city did not see a cost premium in any of the projects that used blended cement. These included two major construction projects, as well as regular use in sidewalk construction.

Local governments routinely purchase concrete for construction projects. Adopting policies for using alternative materials in place of cement is an easy way to reduce greenhouse gas emissions and landfill waste. Local governments can also influence their communities' emissions by supporting the market for blended cement and making information available to the public.

Additional Resources

City of Berkeley Blended Cement Policy Resolution
and Supporting Material
www.ci.berkeley.ca.us/citycouncil/2002citycouncil/packet/121702/2002-12-17%20Item%2027.pdf

Light Emitting Diode Traffic Signals and Exit Signs

Light emitting diodes (LEDs) provide a cost-effective alternative to the incandescent bulbs traditionally used in traffic signals, exit signs, and other specialty lighting applications that local governments are responsible for maintaining. LEDs can consume 80% less energy and last five times longer than conventional incandescent lights.[46] A standard 150-watt incandescent bulb can be converted to a 12-inch LED fixture that uses just 18 watts with no loss of light output.

Lower energy and maintenance costs have proven to local governments that converting to LED technology in traffic signals has a short payback period on the initial investment and saves the jurisdiction money over the lifetime of the light. Currently, red LED signal lights are very cost effective and in common use, but historically the technologies required to produce other colored lights was not as effective. Today, it is common to find green lights in widespread usage as well as white LEDs in pedestrian "walk" indicators. Yellow is the least common, because yellow lights are on a relatively short length of time (compared to red and green), making it harder to justify on the basis of energy savings.

The City of Philadelphia, Pennsylvania, has installed red LED traffic lights in all 2,900 intersections in the city. These energy-efficient lamps produced an annual savings of $800,000 and had four-year return on investment.[47]

Table 14. **Reducing Emissions and Energy Use through LED Lighting**

Philadelphia's LED Traffic Signals	Overland Park's LED Exit Signs
Energy use cut by 83%, saving 64 million kWh annually	Electricity savings of 41,000 kWh
CO_2 emissions reduced by 41,490 tons	CO_2 emissions reduced by 35 tons
Maintenance requirements reduced by six times	Energy savings of $2,750 annually
Energy savings of $800,000 annually	Energy-efficient lighting

SOURCE: Excerpted from International Council for Local Environmental Initiatives, *Best Practices for Climate Protection: A Local Government Guide.*

Overland Park, Kansas, replaced 2,023 red traffic signals with LED fixtures. With annual savings of over $160,000, the LED investment will be paid back in less than eighteen months. These savings do not take into account the extended life of LED bulbs, which can last up to seven years. Traditional incandescent bulbs would have to be replaced multiple times over that timeframe, which incurs ongoing maintenance costs and traffic congestion.

The United States has an estimated 100 million illuminated exit signs in operation twenty-four hours a day, 365 days a year. These signs use over 30 million kWh of electricity annually at a cumulative cost of $1 billion.[48] Older exit signs operate on incandescent light bulbs, but newer models (and retrofit kits) use LEDs. Following on the success of their LED traffic signal program, the City of Overland Park has replaced the exit signs in municipal buildings with LEDs, saving 41,000 kWh and $2,750 annually.[49] Again, reduced maintenance costs are not figured into these savings. The staff time to change out incandescent bulbs repeatedly over the life of an LED exit sign often can cover the upgrade expense even without considering energy savings.

Additional Resources

Light Emitting Diodes for Traffic Signals
Public Technology, Inc.

1-800-PTI-8976
www.pti.nw.dc.us

Public Transportation Promotion

Traffic congestion, parking shortages, and air pollution affect many downtown city centers. The City of Ann Arbor, Michigan, took an innovative approach to get downtown people out of their cars to alleviate some of these
issues.[50] In 1999, Ann Arbor created a program entitled "getDowntown,"
which offers discounted bus passes, the "go!pass," for employees who work
downtown. The program, which was funded by a Congestion Mitigation and
Air Quality Improvement grant, provided unlimited-use bus passes for all
downtown employees. During the two-year pilot, a total of 11,300 go!passes
were distributed to over four hundred downtown businesses. Over a third of
employees who participated in the program increased mass transit use, and
one in ten reduced single-occupancy vehicle use. Based on the success of
the pilot project, Ann Arbor's Downtown Development Authority began
funding the getDowntown program permanently in 2001. Follow-up studies
indicated that not only was mass transit use up, but parking demand was
reduced as well.

The getDowntown program now offers the go!pass to all downtown businesses at the cost of $5 per employee (all full-time employees in an organi-
zation must be provided go!passes), with the remaining cost ($48.21 in 2006)
subsidized by the Downtown Development Authority. Over three hundred
businesses in downtown Ann Arbor currently provide go!passes to almost
five thousand employees. In 2003, the Ann Arbor Transportation Authority
recorded 293,624 boardings by go!pass users, preventing 1,468,120 potential
pounds of CO_2 from being released into the atmosphere.

In addition to offering go!passes, the getDowntown program serves as a
resource to downtown businesses and their employees on other commuting
options such as biking, walking, inline-skating, and vanpooling. The program
also hosts a month-long campaign in May known as Curb Your Car Month,
during which commuters are urged to be creative when choosing a mode of
transportation for getting to work or to the store. During Curb Your Car
Month, local businesses and organizations participate in the getDowntown
Commuter Challenge, a friendly competition to see which business or organization can motivate the highest percentage of its employees to use sustainable modes of transportation during their daily commutes.

In May 2005, the getDowntown Commuter Challenge logged a total of 81,298 alternative transportation miles, thereby preventing approximately 78,000 potential pounds of CO_2 from entering the atmosphere.

Additional Resources

getDowntown Program
Ann Arbor, Michigan
www.getdowntown.org

Chapter Seven

Building a Climate Action Plan
The Experience of Fort Collins

The things we can do to reduce greenhouse gases have other economic and environmental benefits, too, and can help preserve and even improve the quality of life in our communities.[1]

Ray Martinez, Mayor of Fort Collins, Colorado

We have walked the talk and been a leader, Fort Collins is a community interested in improving local air quality, protecting the environment, protecting the health of our citizens, and maintaining a good quality of life."[2]

Ray Martinez, Mayor of Fort Collins, Colorado

Fort Collins, Colorado, is a moderately sized college community of 127,000 people located along the front range of the Rocky Mountains in Colorado, approximately 70 miles north of Denver. It is a regional shopping and employment center for northern Colorado and Southern Wyoming. As such, the city deals with issues similar to other communities of its size. It has been experiencing an average growth rate of 3% and, according to the Chamber of Commerce, has been ranked the sixteenth-hottest housing market in the United States. Associated with this growth, the city finds itself addressing transportation needs, demands on its infrastructure, and continued pressures on the local environment. The City of Fort Collins has been proactive in its attempt to address the various competing issues it faces in a comprehensive fashion. Whether it is through the sustainability inventory and action planning process, the city's Air Quality Action Plan, or its climate-protection initiatives, the city has realized that environmental issues do not exist in a vacuum, but rather are linked closely to the community's social and environmental objectives. Therefore, actions to combat global climate change must involve the whole community and can be a critical component to advancing multiple goals and objectives of the community.

In 1997, the City of Fort Collins passed a resolution officially recognizing global warming as an issue on which the local governments can have a large impact through influencing the communities' energy usage by exercising "key powers over land-use, transportation, building construction, waste management, and, in many cases, energy supply and management."[3] This resolution also recognized that actions taken by the city to reduce greenhouse gas emissions would have multiple community benefits including "decreasing air pollution, creating jobs, reducing energy expenditures, and saving money for the City government, its businesses and its citizens."[4]

This resolution further committed the city to developing a local action plan that would outline the steps the city would take to reduce greenhouse gas emissions from its internal operations and throughout the community as a whole. This plan was to incorporate an initial audit of the quantity and sources of greenhouse gas emissions being released by the community, a forecast of predicted future emission levels, and recommendations for a specific greenhouse gas emission reduction target. Additionally, the plan was to include a strategy for meeting the reduction target, including specific steps to be taken and their estimated impact.

The city completed the initial commitment to climate protection with the adoption of the "Fort Collins Local Action Plan to Reduce Greenhouse Gas Emissions" in the fall of 1999. But rather than marking the end of the city's action, the adoption of this plan was the first step in a much longer journey toward achieving their goals. The action plan adopted by the city council contained a goal to reduce greenhouse gas emissions 30% below the predicted emissions level of 2010. Since its adoption, the Natural Resources Department has been hard at work implementing the policies and programs laid out in the plan and then monitoring progress through biannual assessments of the status of the campaign.

The Action Plan

In designing their Climate Protection campaign, Fort Collins chose to proceed with a "no regrets" approach that emphasizes making economically sound choices to curb greenhouse gas emissions while providing multiple benefits to the community and supporting existing community goals. Throughout development of the Climate Protection Plan, the city emphasized a consensus-building process and actions that achieve multiple benefits for the community. In this way, planners settled on actions that would make sense to implement even if global climate change was not a concern. That is to say, the actions selected also decrease air pollution, create jobs within the community, reduce energy expenditures, improve the quality of life in the region, and/or ultimately save the government, businesses, and citizens money through improving operational efficiency.

The Motivation

As mentioned previously, taking action to reduce greenhouse gas emissions helps Fort Collins act as a responsible environmental steward while meeting multiple community goals. In their 2000 article, "Climate Protection: Fort Collins Likes the Idea," John Fischbach, Fort Collins' City Manager, and Lucinda Smith, Fort Collins' Environmental Planner, outlined the motivating reasons for the city's decision to take action and develop its climate plan. The ten key motivators they laid out are summarized in the following list:[5]

Ability to Effect Change. Urban areas are responsible for about 78% of anthropogenic greenhouse gas emissions, and local governments can play a vital role in reducing these emissions through their activities in overseeing land-use, transportation, building construction, waste management, and energy use.

Risk Reduction. The city followed the warning of the American Geophysical Union that scientific uncertainty over the details of climate change does not justify inaction by policymakers. This message was echoed in a 1999 letter from 570 local elected officials to the federal government highlighting the significant risks that the severe weather events predicted to accompany climate change pose to local communities.

Leadership. The city's climate initiative provides a concrete way for Fort Collins to lead by example, to send a message to community leaders on the importance of emissions reduction, and to advance the city's master plan policy to increase energy efficiency and use of renewable energy sources.

Spotlight Existing Programs. Recognizing the importance of existing energy-efficiency and waste-reduction programs, which make up half of the plan's reductions, was important in driving the plan's adoption.

Key Opportunities for Cost-Savings. Energy-efficiency investments have the potential to deliver significant long-term savings for the local government, businesses, and citizens.

Economic Stimulation. Climate protection can stimulate economic development, especially in the area of green building products and services.

Increased Partnership with Business. The initiatives outlined in the Climate Protection Plan support the master plan policy of developing partnerships that will improve air quality and act as a resource to help the community implement pollution-prevention strategies (see the Climate Wise program case study at the end of this chapter).

Environmental Benefits. Measures taken to reduce greenhouse gases also impact other criteria air pollutants—the five largest measures in the city's plan have been estimated to reduce over 3,300 tons of CO, PM_{10}, SO_2, and NO_x annually by 2010. Additionally, recycling and waste reduction extends space in the community landfill, which is nearing capacity. Finally, the measures outlined in the plan conserve water, lower toxicity levels, and provide other environmental benefits.

Emissions Trading and Reporting. Quantifying greenhouse gas emissions and reductions positions the city to take advantage of the voluntary emissions reporting and trading programs that exist and will prepare them if mandatory programs are ever instated.

Heightened Competitiveness for Grant Funding. Climate protection is an additional avenue through which the city can gain access to grant opportunities to advance environmental programs.

The Process

There are as many ways to develop a local climate-protection plan, as there are cities and counties in the United States. The City of Fort Collins chose an inclusive strategy, gaining support from the beginning from city departments, staff, and the community.[6] In order to ensure support at the municipal level for the Local Action Plan, a Project Advisory Committee was established to oversee development of the plan. Getting buy-in for the concept of a climate-protection strategy and participation in developing the plan at this early stage in the process was a critical key to its success. The Project Advisory Committee therefore included representatives from city departments and citizen boards of the City Council that would be elemental in designing and implementing the plan, such as building and zoning, facility services, fleet services, forestry, natural resources, parks and recreation, purchasing, transportation planning, travel demand management, and utilities.

The city invited local businesses, environmental organizations, and scientists to participate on a Citizen Advisory Committee. This committee included representatives from four citizen boards of City Council as well as the Chamber of Commerce, Sierra Club, Colorado State University, and the U.S. Geological Survey. The primary objectives of the committee were to build consensus for the project early on and to ensure coordination with programs already in place that contribute in one way or another to climate protection. Additionally, Fort Collins formed the Staff Technical Committee, which functioned as a direct link to individual city departments to determine if the proposed measures would work with existing policies. Both of these committees oversaw development of the plan and were great sources of ideas on potential new programs and activities to reduce local greenhouse gas emissions. In order to decide which measures to include in the overall plan, a set of criteria were developed to evaluate each suggestion on the basis of technical or logistical limitations, the size of the potential greenhouse gas reduction, cost, and political feasibility.

Each committee conducted a prioritization process for measures in the plan. A lead staff person in the Natural Resources Department provided much of the logistical support for the committees, with assistance from a local pollution-prevention consultant.

Once staff and contractor resources were in place, an oversight mechanism was set up, and buy-in from key departments and other stakeholders was assured, Fort Collins followed a straightforward process to develop its Local Action Plan based on seven concrete steps.

Step 1. Identify and Measure Effectiveness of Existing Greenhouse Gas Reduction Policies and Programs. The consultant gathered information on the advantages and disadvantages of existing programs, plans for future programs, and reasons for retaining or eliminating certain programs. To accomplish this, she personally interviewed appropriate city staff and community members. This generated information on the range and effectiveness of current energy conservation, transportation and solid waste reduction programs, both existing and planned. This analysis considered programs within and outside of city government.

Step 2. Identify New Programs and Estimate Local Feasibility and Cost-Effectiveness. The consultant gathered information on alternatives suggested by the committees for inclusion in its Local Action Plan. This step was more research-oriented, and the consultant used the internet, professional contacts, existing published documents, and her own technical experience to evaluate choices. In order to estimate the potential greenhouse gas emission reductions from alternative new measures, information was collected from cities and counties that had implemented similar policies and programs. The result was a compilation of alternative measures and their local feasibility.

Step 3. Set Reduction Target and Write the Local Action Plan. The committees met regularly to review the information gathered in Steps 1 and 2. The greenhouse gas reduction goal was identified after evaluating which measures could reasonably be accomplished by 2010. The contents of the LAP included:

- baseline inventory and forecast of greenhouse gas emissions,
- greenhouse gas reduction target of 30% below 2010 levels,
- recommended set of strategies for meeting the target,
- budget needs for recommended strategies,
- recommendation for public education and outreach,
- monitoring and evaluation plan for determining progress.

Step 4. Build Public Support for Plan. Key staff and the consultant developed public participation strategies. Public meetings were held to introduce the plan to the public and gather comment. The first public event, held at the beginning of the process, included a presentation on sustainable building practices and an introduction to Fort Collins' participation in the CCP Campaign. Eighteen months later, after the draft plan had been completed, the city hosted another public open house to gather comments. An internal open house for city employees was held as well. At the conclusion of the comment period, the committees reviewed the comments and evaluated which changes were necessary to include in the Local Action Plan.

Step 5. Evaluate Benefits and Costs. Throughout the meetings, the staff ranked their lists of actions according to priority determined by the pubic and staff. This ranking went beyond personal preference and attempted to take into consideration the return on investment for each of the proposed actions. This assessment smoothed the adoption process by addressing council members' questions about initial costs and expected results.

Step 6. Receive City Council Approval of the Local Action Plan. One study session was held with City Council midway through the process to obtain guidance on plan direction. Following completion of the draft plan and public comment, staff presented a resolution to the council for adoption of the plan. The council adopted it largely because of public support as well as because of its success in meeting the intent of previous council direction.

Step 7. Prepare for Implementation of the Local Action Plan. When adopting the plan, City Council established a municipal Energy Management Team to develop an implementation schedule for measures contained in the plan. Budget requests were made to begin implementing the plan in the next budget cycle. The adopting resolution also called for a biennial report to be submitted to City Council evaluating past greenhouse gas reduction efforts and recommending future measures for consideration beyond those included in the plan.

The Plan

Inventory and Forecast

In order to set an emissions reduction target and develop an effective plan for meeting that target, the city first needed to know the baseline emissions level from which to measure progress.[7] The City of Fort Collins chose 1990 as the base year for their action plan. This was a year for which good data was available on energy use, transportation, and waste generation, and which matched the recommendations of the CCP and those embodied in

Figure 1.   **Fort Collin's 1990 Greenhouse Gas Sources (1.36 million tons eCO$_2$)**

SOURCE: City of Fort Collins

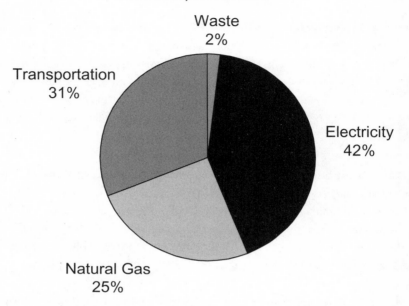

international agreements (e.g., the Kyoto Protocol). This inventory determined that in 1990 the community as a whole was responsible for the emission of 1,365,625 tons of eCO$_2$, or 15.56 tons per person. Electricity use was the largest source of greenhouse gas emissions but overall emissions levels were distributed fairly uniformly among sectors of the community.

The base year's emissions inventory then was used to predict what emissions levels would be in 2010 if no actions were taken to reduce greenhouse gases in the community. This forecast was made by applying annual growth multipliers to the electricity and natural gas usage in 1990 as well as to the transportation fuel used and waste generated. These growth multipliers were supplied by the local utilities, Department of Transportation, and City Planning Department. This analysis determined that the 2010 emissions would be approximately 3,523,000 tons of eCO$_2$ or 159% higher than the 1990 levels. This translates into 24.56 tons per person.

A similar analysis was applied to the city's internal operations. The local government itself produced 39,736 tons eCO$_2$ in 1990 or 2.9% of the overall community emissions. Quantifying the municipal emissions separately helps the city take actions specifically aimed at reducing municipal emissions and meet the directive of the City Council in Resolution 97-97: "The Council intends for the City to take a leadership role in increasing energy efficiency and reducing greenhouse gas emissions from municipal operations."

Target and Reduction Strategies

Building from the inventory of greenhouse gas emissions, the City of Fort Collins complied the emissions-reduction strategies that it wanted to include in its Local Action Plan. This included activities that could be undertaken by departments and divisions throughout the local government and that the city felt it could influence the community to undertake. The list included a combination of existing measures that already had been implemented, pending measures that had not been adopted but would be considered regardless of their impact on greenhouse gas emissions, and new measures that emerged through the process of creating the climate action plan.

Figure 2. **Predicted Future Growth in Greenhouse Gas Emissions and the Impact of Municipal Emissions Reduction Measures**

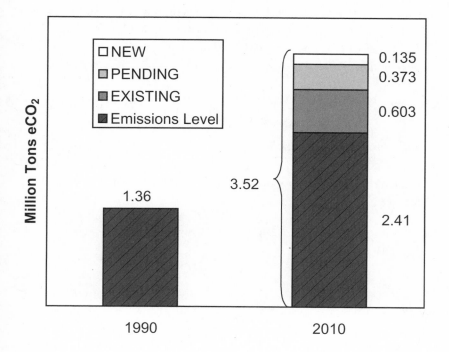

Taken in their entirety, the measures identified for adoption would yield a 32% reduction in greenhouse gas emissions below the forecasted 2010 emissions level.[8] This analysis showed in detail how the city could influence its emissions levels and was the key factor in the city's decision to officially adopt a 30% reduction target that would limit the community's emissions growth to 77% from the 159% originally predicted.

Sample Measures in the Plan

Transportation

SUPPORT FORT COLLINS-DENVER RAIL.

Strong citizen support exists for the development of a light-rail line between Fort Collins and Denver.[9] Fort Collins' Local Action Plan recommends building local transportation infrastructure to accommodate or improve access to potential future rail links. The plan estimates that 15,000 to 50,000 tons of CO_2 would be reduced from the construction of a rail link.

PROMOTE TELEWORKING.

In addition to developing an internal city policy supporting telecommuting, the Transportation Department conducts outreach to local businesses, providing information about the benefits of telecommuting. Approximately 3,076 tons of CO_2 will be eliminated in 2010 if 5% of citizens telecommute twice a month.

Energy

REPLACE TRAFFIC SIGNALS WITH LEDS.

The City of Fort Collins is replacing incandescent traffic signals with LEDs, which consume less energy, last longer, and require less maintenance. Converting all intersections to LEDs could save the city 3,137 tons of CO_2 and $101,961 each year.

INCREASE USE OF WIND ENERGY THROUGH
GREEN PRICING PROGRAMS.

About 10,256 tons of CO_2 will be saved yearly by 2010 by encouraging, facilitating, and regulating energy efficiency and the use of renewable energy resources. This process will involve both the public and private sectors and will involve information and educational services, financial incentive programs, requirements and incentives in the planning process, and enforcement of regulations such as the Energy code.

Solid Waste

ACHIEVE 50% WASTE DIVERSION BY 2010.

The city will achieve an 112,787-ton CO_2 savings by 2010 through the expansion of the existing recycling center, the construction of a central drop-off site, and expanding public outreach efforts on recycling.

Vegetation

PLANT TREES CITYWIDE.

About 125 tons of CO_2 will be saved by 2010 by offering matching funds to nonprofits to plant trees, paired with an education campaign promoting tree planting.

Purchasing

BUY EFFICIENT FLEET VEHICLES.

This program will provide an aggressive education campaign encouraging city departments to purchase smaller, fuel-efficient vehicles.

DEVELOP A "GREEN GUIDE" FOR ENVIRONMENTAL PRODUCTS.

Fort Collins is developing a guide for purchasing environmentally preferable products, to be distributed to Fort Collins residents.

Education and Outreach

DEVELOP GLOBAL WARMING TEACHER KITS FOR SCHOOLS.

This measure will be modeled after other successful public education curriculums, such as that used in Chula Vista, California. The teacher kits will include lesson plans that discuss greenhouse gases, global warming, and energy efficiency.

Results to Date

Since the plan was adopted, the city's Energy Management Team has been working on its implementation. They are also tasked with preparing a biannual report on the city's progress and recommending future actions.

Table 15. **Fort Collins Emissions Reductions (In Tons)**

	2003	**2004**
Citywide	186,000	213,000
Municipal	33,000	28,000
Total	**219,000**	**241,000**

SOURCE: City of Fort Collins Energy Management Team, "City of Fort Collins 2003/2004 Climate Protection Status Report," 2005.

The most recent of these reports was released in the fall of 2005. In this report period, they reported that emissions levels in the city increased between 1990 and 2004 by 81% to 2.467 million tons.[10] At the same time, it was estimated that the measures the city had put into place were avoiding approximately 241,000 tons of eCO_2 annually, which was 9% of the community's total emissions for the year. Citywide emissions levels would have been closer to 2.71 million tons if no action had been taken on the part of the city.

Although it is still early to predict with certainty whether the city will meet its 2010 reduction target, progress is being made. To date, Fort Collins has succeeded in holding emissions growth fairly constant and at a level that will achieve its targets. The city's action plan and its updates continue to provide an outline for emissions reduction measures still to be implemented. With a viable roadmap and the political will to continue, the city should be able to continue moving toward meeting its target for fighting global climate change.

Future Success?

As they entered the process, the City of Fort Collins acknowledged that achieving its goal of a 30% reduction in emissions below those predicted for 2010 would be a challenge that would require an array of new and innovative measures to be implemented. To do so will require the support of all segments of the community to turn their collaborative planning process into a communal commitment to turning the plans into actions and results.

Figure 3. **Fort Collins—Historic Emissions Levels and Target**

SOURCE: City of Fort Collins Energy Management Team,
"City of Fort Collins 2003/2004 Climate Protection Status Report," 2005.

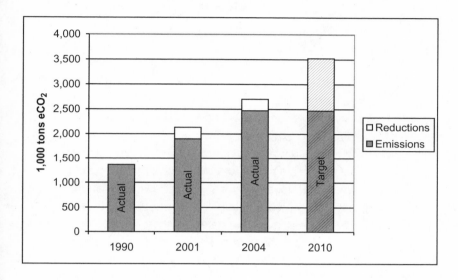

The municipal government needs to show on-going support in the form of leadership and to provide for the allocation of staff and financial resources to ensure that the programs and initiatives remain on track. Together, the business and industrial sectors are responsible for 43% of the community's greenhouse gas emissions. Therefore, for Fort Collins' Climate Action Plan to succeed, it will take action on the part of the entire business community actively accounting and working to reduce their energy use and waste generation. Finally, it will take the commitment of all citizens to ensure the success of the policies and programs put into place. Only through individual actions can the jurisdiction effectively reduce the scope of their transportation, residential, and waste emissions—which together account for over half of the city's climate-change footprint.

Fort Collins has followed through on a world-class process for emissions reductions that has set them on course toward meeting their target with the support of the community, but it is now up to all sectors to follow through in support of actions needed to achieve those goals.

Case Study: Reaching Out to Local Businesses—The "Climate Wise Program"

Local governments have a great deal of control over the emissions from their internal operations, but ultimately a successful emissions reduction program requires reaching out to the community as a whole. In working to achieve a community's emissions reduction goals, local businesses can be strong allies. Not only does the commercial sector account for approximately 43% (combined commercial and industrial sectors in the City of Fort Collins) of the community's overall emissions, but local businesses meet and interact with a broad cross section of the population, making local businesses a powerful gateway to the community. Reaching out to business is often one of the first actions that a government takes after addressing its own internal operation. Despite the potential for emissions reduction, environmental protection is often not a top priority for business owners and managers who have to balance many interests that compete for their time and resources.

The City of Fort Collins has been particularly successful in reaching out to the business community through its award-winning Climate Wise Program. This innovative initiative creates a one-stop gateway for a number of related pollution-prevention services available to local businesses—from energy efficiency and green power to solid waste reduction and transportation demand management. The Climate Wise program is voluntary. Businesses that participate receive technical assistance, public recognition, and networking opportunities. In exchange, the businesses commit to taking steps to actively reduce their emissions of greenhouse gases.

The Climate Wise program was launched in 2000 as a key component of the emissions reduction program. In Fort Collin's original Climate Action Plan, it was anticipated that this program could eliminate 38,390 tons of greenhouse gas emissions by 2010.[a] Since its inception, over fifty local establishments have become participants in the Climate Wise program. According the city's 2005 status report on climate change activities, actions taken by these businesses have reduced greenhouse gas emissions by 94,746 tons, 2.5 times the original target.[b]

The success of the Climate Wise program lies in the fact that it addresses the issues businesses are concerned with and is not billed strictly as a climate-protection initiative. Through the Climate Wise program, the City of Fort Collins helps businesses tackle challenges that impact bottom lines and the quality of life in Fort Collins. Businesses that join the program are provided with free technical assessments of their energy and water usage, waste generation, and transportation demand. They also receive public recognition and publicity of their successes and various opportunities throughout the year to network with their peers.

Technical assistance is provided either by members of the Climate Wise team, by the businesses themselves through resources made available by the city, or through partner agencies. Assistance provided includes:

- An *on-site assessment* to identify specific steps that the business can take to reduce greenhouse gas emissions. The assessment focuses on energy efficiency, pollution prevention, solid waste reduction, or travel reduction.

- Providing advice on *project selection, action planning,* and *implementation*. Although the final decision on which actions to undertake is ultimately up to the businesses, the Climate Wise program can provide suggestions and advice. The Climate Wise team has expertise in a variety of fields and is available as a resource for general questions or technical assistance throughout the planning and implementation stages.

- Help in developing systems for *monitoring progress* and *submitting annual progress reports* through in-person assistance and on-line forms.

- Inclusion in *networking events* with peer businesses to exchange ideas and successes.

In addition to the technical assistance that businesses that join Fort Collins' Climate Wise Program can receive, they also may gain free public recognition for their commitment and progress toward reducing greenhouse gases. This includes:

- decals to post in the business to inform people that the business is a participant;

- official recognition from the mayor and City Council;

- participation in an annual Recognition and Awards Event that publicly acknowledges the accomplishments made;

- public recognition through advertisements run in local papers thanking the participating businesses and news articles or press releases publicizing the success with greenhouse gas reduction.

In 2003, Fort Collins received national recognition for their Climate Wise Program. In a ceremony at the U.S. House of Representatives Office Building, in Washington, D.C., the city was given the Most Valuable Pollution Prevention (MVP2) award from the National Pollution Prevention Roundtable. This is a prestigious (and highly competitive) award given to government agencies, non-profit organizations, or industries who have taken significant steps in the area of pollution prevention.

---

a. City of Fort Collins Local Action Plan.
b. City of Fort Collins Energy Management Team, "2003/2004 Climate Protection Status Report."

Chapter Eight

Achieving Results
Portland, Oregon

> Portland and cities throughout the world are responsible for creating a sustainable future for our children. We know that cutting CO_2 emissions is not only smart for the environment, it's great for business, too. If we reduce our CO_2 emissions, we also reduce local air pollution, plant more trees, lower energy bills for residents and business, use more solar and wind power, and create a more livable, walkable, community-oriented city for all of us. Cities must take a leadership role. We cannot wait for federal action.[1]
>
> Vera Katz, Mayor of Portland, Oregon

The City of Portland, Oregon, was an early local government adopter of the challenge to reduce local greenhouse gas emissions. Portland developed and adopted its climate action plan in 1993, becoming the first community in the United States to take official action to address global warming. Being an early adopter has paid off. In the spring of 2005, Portland was able to announce that it has held community-wide emissions to 1% over 1990 levels.[2] This is particularly impressive when viewed through the lens of the significant population growth that the region has undergone in recent years. When viewed on a per capita basis, Portland's emissions have dropped by approximately 12.5% since 1993 when the jurisdiction formally committed to taking action to reduce greenhouse gas emissions.[3] The magnitude of this reduction shows that it is possible for a community to stabilize and significantly reduce its emissions.

Background

The city of Portland is a 500,000-person urban center that makes up the core of a larger 1.7 million-person metropolitan region. Situated at the confluence of the Willamette and Columbia rivers, and under the shadow of Mount Hood, Portland is considered to be an attractive, hip, environmentally oriented city, and was ranked as the "Best Big City" in the United States by *Money* magazine in 2000. But this was not always the case.

In the 1950s and 1960s, the city was overtaken by freeways that fueled urban flight to the suburbs. In order to accommodate America's reliance on the automobile, downtown began to change, with older buildings being replaced with parking lots to accommodate commuters. The loss of population from the urban core was countered by various "urban renewal projects" that developed high-rise apartment complexes in areas that formerly were dominated by vibrant interconnected neighborhoods.

Fortunately for the city, this kind of depopulation, which has lead to the death of the urban community in many large cities around the country, was stemmed by visionary planners who began to appear on the political scene in the early 1970s. Decrying the urban sprawl that was beginning to dominate the landscape, Governor Tom McCall, during his opening address to the state legislature in 1973, announced that the state was, in dire need of a state land-use policy, new subdivision laws and new standards for planning and zoning . . . we must respect one truism: That unlimited and unregulated growth leads inexorably to a lowered quality of life.[4]

The state rose to the challenge later that year, adopting Senate Bill 100, which instituted a state-wide planning process to limit sprawl and to protect and preserve natural spaces, forests, and farms. Under this new law, urban areas were to designate urban growth boundaries around existing population centers of sufficient size to accommodate predicted population growth. Planning and zoning decisions within those "urban growth boundaries" were to promote higher density development consistent with an urban community character, while areas outside the growth boundaries were to be managed in ways to preserve their rural characteristics. The effect of this legislation was to begin to focus development pressures on existing urban areas and to begin to curb sprawl.

This policy mirrored the vision of the City of Portland itself, which adopted a new downtown plan in 1972 that placed an emphasis on transit-oriented development and envisioned a vibrant downtown with increased housing and retail opportunities that were active twenty-four hours a day. In line with this philosophy, increasing parking was discouraged in favor of preserving historic buildings, and a new transit mall was constructed to anchor the downtown and provide alternatives to personal vehicle travel. The height of this shift was seen when the Harbor Freeway was removed to be replaced by an urban park along the waterfront and a proposed Mount Hood Freeway was shifted to a plan to meet transportation needs through the development of a light-rail system.

Through these and many other initiatives, the city was able to shift the tides of urban sprawl and lay the foundation for a more sustainable community that, twenty years later, is able to claim that it is on track to meet its emissions reduction goals.

Climate Change Planning

The City of Portland was the first U.S. city to adopt a global warming policy for reducing greenhouse gas. In 1991, it joined 13 other local governments around the world in participating in the International Council for Local Environmental Initiatives (ICLEI) Urban CO_2 Reduction Initiative.[5] This initiative brought together American, Canadian, and European cities at six working meetings to develop a municipal planning framework for greenhouse gas reductions and strategic energy management.

As part of this initiative, members of Portland's City Council and various staff members from the planning and other departments were involved in technical exchanges with their counterparts in Stockholm and Copenhagen. Through this experience, the city was able to develop a plan drawing on the knowledge of their European colleagues who had a greater range of understanding in taking action to reduce greenhouse gas emissions. Progressing through this framework gave rise to Portland's emissions reduction strategy, which was adopted shortly after the conclusion of the Urban CO_2 Reduction Initiative in 1993. Portland's plan adopted a goal of reducing greenhouse gas emissions by 20% below 1988 levels by 2010.[6]

In 2001, Multnomah County joined the city in committing to reduce greenhouse gas emissions, and the city's plan was expanded to take a more regional approach. By combining efforts, both jurisdictions recognized that climate change and its mitigation was not the purview of any one geographic entity. Real emissions reductions require a coordinated effort. Without this, there is a risk of lowering emissions in one location by exporting them to another. By taking a unified approach, Portland and Multnomah County help to ensure that actions taken truly reduce emissions throughout the region.

At the same time that the city and county combined efforts, the city adjusted its emissions reduction goal. The new target was to reduce greenhouse gas emissions by 10% below 1990 levels.[7] This new goal was taken out of recognition that the rapid population growth experienced by the region would make the original, more-stringent target difficult, if not impossible, to meet. Although this was a lowering of the standard, it is still more aggressive than those set by the Kyoto Protocol for the United States. Additionally, the 2001 revision to the plan stated that this goal is only a beginning. If the Portland region is able to maintain a trajectory of reducing greenhouse gas emissions by 10% every twenty years, it will come close to achieving the 60 to 70% reduction needed to stabilize concentrations of greenhouse gases by the mid- to end of the century.[8]

The Actions

The city's Office of Sustainable Development and county's Sustainability Initiative are working together to implement their Local Action Plan on Global Warming. As in the City of Fort Collins, the Portland/Multnomah County Plan came out of a participatory public process that brought together citizens, businesses, nonprofit organizations, the utilities, and local government agencies. These groups met together in formal discussions throughout the summer, and over three hundred copies of the draft plan were distributed for comment in the fall of 2000, with multiple follow-up meetings held between local governmental departments and various constituency groups.

The plan contains more than 150 actions for implementation along with specific timelines for when each is to go into effect. The action items can be organized into six major focus areas, detailed in Table 16.

In order to meet these goals and in turn make progress toward the overarching 10% reduction target, the city and county governments have been taking a number of innovative actions in each of the above thematic areas. Samples of the actions included in the plan and that have been implemented are included below.

Policy, Research, and Education. The plan contains a strong informational component so that the best information is available to guide action and inform policy. Elements of this component include:

- completing a biannual report of local greenhouse gas emissions levels to monitor changes and adjust the plan accordingly;

Table 16. **The Focus of Portland's Emissions Reduction Plan and Anticipated Reductions**

Focus Area	Reduction Target (millions of metric tons)
Providing *policy, research,* and *education* for local agency staff and throughout the community as a whole	n/a
Increasing *energy efficiency in buildings* to reduce energy use in facilities across all government and community sectors by 10%	0.67
Promoting options for *transportation, telecommunications,* and *access* to reduce per capita vehicle miles traveled by 10% below 1995 levels and to improve the average fuel efficiency of vehicles from 18.5 to 26 mpg	1.35
Emphasizing *renewable energy resources* to meet all growth in electricity load through renewable sources	0.54
Promoting *waste reduction and recycling* to reduce solid waste generation and improve recycling rates to minimize the methane emissions from area landfills as well as manufacturing processes	0.23
Enhancing *forestry* and *carbon offsets* through expanding urban and rural forestry practices and actively seeking other opportunities for acquiring carbon offsets	0.31
Total Estimated Reductions	3.10

SOURCE: Portland Office of Sustainable Development, Multnomah County Department of Sustainable Community Development, "Local Action Plan on Global Warming: City of Portland and Multnomah County," 2001.

- interoffice communications and training sessions on the challenges global warming poses to the community and the government's strategy for emissions reduction;

- Community outreach and education programs conducted by the City, neighborhood associations, advocacy groups, religious organizations and local businesses have educated their members, and

- Advocacy for national action on global warming.

Energy Efficiency in Buildings. Portland's plan seeks to provide cost-effective methods to improve the efficiency of structures within the region while supporting community values, local businesses, active and healthy neighborhoods, and especially assisting low-income residents. To that end, the plan provides:

- investment in all energy-efficiency measures that have simple paybacks of ten years or less;

- public-private energy-conservation partnerships;

- adoption of green building standards for municipal construction and offering green building technical assistance, education, and financial incentives to private builders who meet Leadership in Energy and Environmental Design (LEED) standards;

- conversion of traffic lights from incandescent to LED bulbs; and

- countywide energy-use best practices performance and standards for equipment, lighting, heating and cooling, appliances, and personal computers.

Transportation, Telecommunications, and Access. Almost 40% of Portland's emissions come from the transportation sector, which is a key area of interest in the plan. The plan focuses on:

- providing education on rebates, tax credits, and other incentives offered by the State of Oregon;

- Portland's purchase of over thirty hybrid vehicles since 2001;

- city and county switch of diesel vehicles to biodiesel blends;

- county fuel-efficiency standards set for new vehicle purchases and conversion of its diesel fuel to biodiesel;

- developing more than 200 miles of bikeways;

- actively promoting density, infill, and mixed-use development;

- opening three major light-rail lines; and

- launching the first modern streetcar line in the United States.

Renewable Energy Resources. The city also seeks to encourage the use of energy sources that do not produce climate-changing emissions or deplete natural resources, including

- development of renewable energy demonstration projects such as small-scale solar at parking pay stations;

- Portland's construction of a fuel-cell electricity generator powered by methane from the city's wastewater treatment plant, which produces enough electricity to power 120 homes;

- The city's purchase of 44 million kWh of wind power (through green tags) from the Stateline Wind Energy Center in eastern Oregon and Washington; and

- installation of solar cells on maintenance vehicles so that the engine does not have to be left idling to run power tools.[9]

Waste Reduction and Recycling. The emissions reduction program focuses on the waste to reduce the use of raw material, improve the energy efficiency of the manufacturing process through use of recycled materials, and decrease the emissions from landfills. The plan specifies

- the requirement that 50% of solid waste from businesses be recycled;

- a sustainable paper use policy that minimizes waste-paper generation from the city's operations;

- implementation of commercial food-waste collection and recycling programs; and

- investigating and setting standards for purchasing recycled content materials.

Forestry and Carbon Offsets. The plan seeks to take advantage of Portland's natural ecosystems to utilize native tree cover to sequester greenhouse gases at home or elsewhere (through the purchase of offsets) by

- supporting reforestation initiatives, including planting 3,000 acres of trees;

- documenting the benefits provided by urban forest cover and using this information to inform policy decisions and seek funding to support urban forestry initiatives; and

- promoting the planting of tree species to maximize their carbon offsets, energy conservation, air quality, stormwater management, and habitat benefits.

Results

Portland has conducted regular re-inventories of its greenhouse gas emissions since the city adopted its plan in order to monitor how successful it is in meeting its goals. This ongoing evaluation has shown that continued progress has been made. In 1999, Portland's per capita emissions were 3% below 1990 per capita emissions levels.[10] In 2004, the absolute emissions levels for the city and county were only slightly (1%) higher than in 1990 and per capita emissions had fallen by 12.5%.[11] In addition, preliminary figures for 2005 suggest that emissions may have dropped below 1990 levels in that year.[12] In effect, the greater Portland region has been able to hold emission levels constant over fourteen years despite rapid population growth and economic growth of over 60% (see Figure 4).

Beyond the benefits provided in terms of combating global climate change, actions taken to achieve these emissions reductions have many tangible financial benefits as well. The city government is now meeting 11% of "mizing dependence on foreign oil and the ensuing price volatility. Implementing an energy management system is saving the city $2 million annually and is increasing facility energy efficiency by 15%. The city's conversion to LED traffic signals is an action that paid for itself in two years and now saves $400,000 each year in electricity and maintenance costs. Similarly, an apartment effi-

Figure 4. **Multnomah County's Greenhouse Gas Emissions**

SOURCE: Chart adopted from "A Progress Report on the City of Portland and
Multnomah County Local Action Plan on Global Warming; June 2005 with
additional data from Portland Office of Sustainable Development, Local Action
Plan on Global Warming: City of Portland & Multnomah County," 2001.

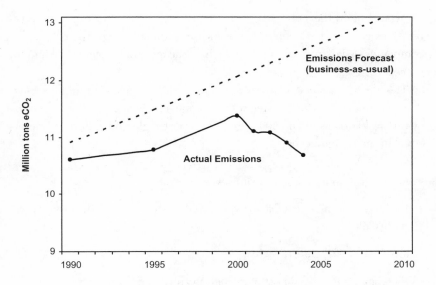

ciency program, funded through grants, is saving residents over $3 million
annually. The cumulative savings for households and businesses participating
in public and private energy-conservation efforts, between 1990 and 2003, are
estimated to exceed $300 million.[13]

Portland's success belies the common mantra that taking action on climate
change is antithetical to strong economic growth. When *Money* magazine declared Portland the "Best Big City" in which to live, they made only passing reference to the region's growing high-tech sector and devoted most of the article
to promoting the aspects of the city that create an atmosphere in which workers become a part of the community and businesses want to locate.[14] These
same traits are the backbone of Portland's success in reducing greenhouse gas
emissions. Characteristics identified in the *Money* magazine article include:

- planning to minimize urban sprawl;

- short blocks that encourage walking and a "café culture";

- ease of mobility on public transit, via streetcar and light rail (which encompasses an airport connection serving both locals and out of town visitors); and

- redevelopment of the barren riverfront.

According to the same article, the culture developed through these actions has encouraged people to live and work downtown to the extent that companies have to contend with a 3% vacancy rate when looking for space.[15]

The city attributes its success to a range of programs, including land-use planning but also encompassing waste, energy efficiency, and the promotion of renewable energy. The 2005 progress report highlights a number factors contributing to its success:

- a 75% growth in public transit use since 1990;

- the purchase of renewable energy for more than 10% of the city governments' electricity;

- a 54% recycling rate;

- the construction of almost forty high-performance buildings;

- the planting of over 750,000 trees and shrubs since 1996;

- the weatherization of 10,000 multifamily units and over 800 homes in two years; and

- the establishment of the Energy Trust of Oregon to provide consistent funding for energy efficiency and renewable energy programs.[16]

The hard work of the City of Portland was recognized by the U.S. Environmental Protection Agency in 2002 with the receipt of the EPA's Climate Protection Award. These awards recognize individuals, businesses, or governments that lead the fight toward preventing global climate change. Winners are selected for taking original action that leads to a significant reduction in greenhouse gas emissions, while showing persuasive moral and/or organizational leadership, and implementing a program with a uniquely global perspective and impact.

In accepting this award, Mayor Vera Katz reiterated the connection between economic growth and environmental protection, stating that, "a healthy environment is key to a thriving economy. Our high quality of life is what makes Portland so appealing to those across the nation. It makes them want to travel here, it makes them want to do business here, it makes them want to raise their families here."[17]

A number of different structural factors have played into Portland's success. The city has been able to make these strides forward due to the comprehensive nature of its action plan and the programs that have been implemented. An effort has been made to reach out to all sectors of the local and regional governments as well as to the community as a whole. Ongoing monitoring and evaluation of their programs has allowed the city both to track its successes and to identify opportunities for improvement. Finally, there is a deep commitment to reducing emissions and an understanding that actions taken to protect the environment also help to improve the quality of life of the local citizenry.

Going Forward

Although Portland has realized a significant achievement, the city has renewed its commitment to continue working toward its 10% reduction goal. Activities slated for implementation in the next few years include:

- increasing action aimed at reducing community-wide emissions—such as expanding the city's composting program to capture commercial food waste and improving the overall recycling rate to boost it to 60% by the end of 2005;

- programs to reduce emissions from the city and county governments' internal operations, including conducting a feasibility study of supplying city facilities' entire energy demand through the use of wind power and implementing a "Transportation Options" campaign aimed at reducing automobile travel by county employees;

- institutionalizing the city's commitment to climate protection by joining the Chicago Climate Exchange—by entering into this voluntary partnership, the city will make a legally binding commitment to reduce emissions from its internal operations. If it fails to meet those reductions, it will have to buy carbon credits to offset the shortfall. This provides an additional financial incentive for making progress;

- enhancing climate protection throughout the region through continued involvement in and support of both the West Coast Governors' Global Warming Initiative and the Oregon Governor's Global Warming Strategy.[18]

According to Abby Young, Director of Strategic Planning at the ICLEI— Local Governments for Sustainability:

> Portland is one of the cities in the world that has the most rigorous greenhouse gas quantification processes. Their local climate action plan, in place since the early 90's, includes many greenhouse gas-reducing activities including use of renewable energy, smart growth style land-use and zoning policies, public transportation, waste reduction, extensive energy conservation and efficiency efforts (including a weatherization program that creates jobs in low income neighborhoods), and tree planting. The methodology that the city uses to quantify reductions in greenhouse gas emissions is based on calculating reductions in fossil fuel use and waste going to landfills. So, they're actually doing quite a fine job at quantification, setting high standards for the rest of the country.[19]

Chapter Nine

Warming to the Future

> The reality of global climate change is urgent. The stakes are high—locally and globally—and we need to act. As a city government, we've already cut our greenhouse gas emissions by more than 60% compared to 1990 levels. But it's not enough—we need to work together as a community to set responsible limits on global warming pollution.[1]
>
> Mayor Greg Nickels, Seattle, Washington

As the above quotation indicates, it is becoming increasingly difficult to deny the reality of global climate change. In February 2007, the International Panel on Climate Changes released the most recent assessment of climate change and its impacts. Among other things, this body reported that:[2]

- CO_2, CH_4, and N_2O concentrations far exceed pre-industrial values and this is primarily the result of fossil fuel use, land-use change, and agriculture;

- osberved warming trends are "unequivocal";

- the increases in globally averaged temperatures since the mid-twentieth century are "very likely" the result of increases in anthropogenic greenhouse gas concentrations;[3]

- the results of this warming include melting of snow and ice, rising sea levels, as well as changes in precipitation, ocean salinity, wind patterns, and extreme weather events;

- stabilizing greenhouse gas concentrations at 2000 levels will lead to approximately 0.18°F warming per decade, while current or increased emissions rates will cause further warming, and even if greenhouse gas concentrations were stabilized, warming trends and sea level rise will continue for centuries.[4]

Despite these dire predictions and in spite of the ambivalence displayed to date by this country's leadership toward the issue of global warming, there is cause for hope. Since 1988, the movement of local governments has grown from one (Toronto) to hundreds of communities around the world and an increasing number of local governments are expressing their desire to tackle the issues brought on by uncontrolled emissions of greenhouse gases every day. Even in the United States, which has steadfastly refused to take federal action, 358 mayors have signed the U.S. Conference of Mayors climate protection agreement and many other local governments have committed to action through the Cities of Climate Protection Campaign and other initiatives aimed at this important sector of society.

Just as there is no single source of greenhouse gases, there is no single solution. Climate change is just as much a local issue with local solutions as it is a global issue. At best, international treaties and national caps can set a framework for action. Ultimately, it will be the combined efforts of many individual initiatives that will begin to shift the way that humans relate to their surroundings: continuing down the unsustainable path of reliance on fossil fuels and the consequences that brings, or modifying our transportation, energy, and waste systems to reduce the amount of climate-changing gases and other pollutants we are releasing into the atmosphere.

Local governments have a great advantage over larger management units in that they tend to be smaller, less bureaucratic, and more flexible than national, international, or even state institutions. They are closer to their constituencies and can monitor directly the effects of policies and actions that are put into place and therefore adjust more readily to changing situations. Local governments can experiment more readily with innovative policies. Other communities can adopt the programs that prove successful, and if an approach does not have the desired impact, changes can be made without the entire country having to change course midstream. Thus, the local level is the ideal level at which to tackle greenhouse gas emissions reductions.

The experiences of the early adopters—the local governments that have been striving to address global warming since the late 1980s—have helped to illustrate some key features that can ensure the success of local climate change programs. To develop a truly successful program, it is important that a local government:

- Have a tangible goal against which progress can be measured. Too often, policies are set with vague objectives, which leads them either to be over-looked or without a mechanism to measure success.

- Develop a formal action plan with specific actions, responsibilities, and performance criteria laid out in it. Building a coordinated effort in which actions and policies support one another greatly improves the chances for success over implementing a collection of actions that are unrelated.

- Tie emissions reductions to other local goals and objectives. As discussed in Chapter Three, climate protection can address communities' public health, economic, and quality of life objectives. Tying emissions reductions to these efforts and integrating them into larger municipal plans will help en-sure longevity in the face of limited resources and competing priorities.

- Involve the public and various sectors throughout the process. Ulti-mately, it is the public that will have to implement most of the policies that are adopted. Therefore, involving affected sectors from the outset, so that they view themselves as stakeholders, will improve the chances for a program's success.

- Form partnerships within and among communities. The greater number of resources, skills, and points of view that are brought to the table, the better. It is quite likely that other local programs can support emissions reductions. Facilitate systems where different entities can contribute in their areas of expertise.

- Monitor actively the outcomes and results. No matter how well programs are planned ahead of time, unexpected outcomes can occur. The imple-menting agency should be aware of the direct and collateral impacts that its programs have.

Most importantly, local jurisdictions must be flexible and willing to inno-vate. If a monitoring regime shows that a program is not having the desired outcome, the local government should have the flexibility to adjust. Addi-tionally, because climate change is a larger problem than local communities regularly tackle, solutions often will require "outside of the box" thinking. Local governments should be open to innovations rather than adhering to business-as-usual policies.

This emphasis on local action is not to say that state, federal, and international policies are not needed as well. Federal and international regulations fill a vital niche in the policy arena. They are needed to provide an overall framework in which local government action can be effective. A model for this exists with the Clean Air Act. The federal government sets the overall air pollution standards, but it is up to local regions to develop their own air pollution control plans to ensure they remain in compliance with those standards.

Similarly, a national policy will help ensure a level playing field so that local governments that do adopt innovative policies will not be discriminated against. Rather, all stakeholders know that they will be subject to similar constraints on climate-changing emissions wherever they locate. (A very similar case can be made for taking part in international treaties—in that way, no single country puts itself at a competitive advantage or disadvantage.) Additionally, the larger the scope of activity, the greater the likelihood of finding cost-effective solutions. For example, if more entities fall under a cap-and-trade program, a much greater opportunity exists for market forces to ensure that the least-cost alternatives are followed.

Finally, the federal government also can be a source of resources and innovation. Through its grant programs and the technical and policy assistance of agencies such as the EPA and Department of Energy, the federal government can advance local policies. The national laboratories and research institutes can help develop new technologies and solutions. And when local polices are failing or action is not occurring, the federal government has the resources to encourage action to occur (for example, in the case of the Clean Air Act, the federal government can withhold highway funding for communities that are out of compliance with air quality standards and do not have a mitigation plan in place.)

What federal and international policies should avoid doing is micromanaging and adopting specific strategies that must be implemented at the local level. The federal government can encourage, suggest, and assist, but ultimately, local governments will know best how to develop policies and programs for their communities. Rarely does a one-size-fits-all policy work in a geographically, culturally, economically, and demographically diverse country.

Currently, this level of intervention is not the chief concern of the United States. The United States finds itself without any strong leadership or direction on the federal level to combat global climate change. Despite this lack of federal leadership, a great deal of action is taking place—at the local level. Throughout the country, cities and counties are implementing local greenhouse gas reduction plans, and exerting pressure on the federal authorities. States and cities are taking a leadership role in forcing federal action, even suing to have carbon dioxide regulated as a pollutant, a position recently supported by the U.S. Supreme Court over the administration's objections. Regions (the Pacific Northwest, the Northeast) are forming their own market-based solutions to address climate change with a cap-and-trade mechanism, which will provide an interesting feasibility study for the applicability of such measures across the nation, and indeed, the world. Every day, local governments are experimenting with new policies to reduce emissions of carbon dioxide and other greenhouse gases.

Hopefully soon the United States will take action to curb global warming emissions, whether the motivation comes from external pressures or because the impacts of excess carbon dioxide in the atmosphere have become undeniable. When the federal government does act, it will not have to start from scratch. Policymakers will be able to draw on the experiences of local governments that have been taking action to reduce emissions for almost twenty years.

Will these measures address the global warming challenge fast enough to make a difference? That is the "million-dollar" question. Humanity is being forced to address an issue that has never been experienced before—the long-term results are solid theories, but no one really knows how great or little the extent of damage from a rapidly warming planet will be. But it seems smart to be on the safe side. Reducing greenhouse gas emissions has so many ancillary benefits that it makes sense to do it if only in order to reduce air pollution, save money, and create a higher quality of life. Even if a rapidly changing climate does not result in the extinction of thousands of species, melting of polar ice caps, rising sea levels, and intensification of natural disasters, the fiscally, socially, and environmentally prudent course of action is to take immediate steps to reduce the copious amounts of greenhouse gases being emitted into Earth's atmosphere every day, month, and year.

The more than two hundred local governments in the United States that are participants in the Cities for Climate Protection campaign already have prevented millions of tons of eCO_2 from reaching the atmosphere, and they represent only 19% of the country's population. It is crucial—and more important, possible—to get the remaining 81% involved in the process and working toward a sustainable future for our children and our planet.

Emission Reduction Targets under the Kyoto Protocol

The Kyoto Protocol set emissions limits on countries that sign on. These are measured in terms of the percent change from 1990 emissions that are allowed by the end of the first reporting period in 2012. The table below lists the emissions targets set forth by the protocol. A negative number indicates the amount that emissions must be reduced before 2012. A positive number indicates that emissions are allowed to increase under the Kyoto Protocol.

Required Reductions Under the Kyoto Protocol

Country	Percent Change	Country	Percent Change
Australia	8%	France	−8%
Austria	−8%	Germany	−8%
Belgium	−8%	Greece	−8%
Bulgaria*	−8%	Hungary*	−6%
Canada	−6%	Iceland	10%
Croatia*	−5%	Ireland	−8%
Czech Republic*	−8%	Italy	−8%
		Japan	−6%
Denmark	−8%	Latvia*	−8%
Estonia*	−8%	Liechtenstein	−8%
European Community	−8%	Lithuania*	−8%
		Luxembourg	−8%
Finland	−8%		

(continued)

Country	Percent Change	Country	Percent Change
Monaco	−8%	Slovenia*	−8%
Netherlands	−8%	Spain	−8%
New Zealand	0%	Sweden	−8%
Norway	1%	Switzerland	−8%
Poland*	−6%	Ukraine*	0%
Portugal	−8%	United Kingdom of Great Britain and Northern Ireland	−8%
Romania*	−8%		
Russian Federation*	0%		
Slovakia*	−8%	United States of America	−7%

*Countries that are undergoing the process of transition to a market economy.

SOURCE: Kyoto Protocol to the United Nations Framework Convention on Climate Change, www.unfccc.int/essential_background/kyoto_protocol/items/1678.php (accessed on March 31, 2007).

World Mayors and Municipal Leaders Declaration on Climate Change

Fourth Municipal Leaders Summit on Climate Change
On the Occasion of the United Nations Climate Change Conference
(COP-11 and COP/MOP 1)
7 December 2005, Montreal, Canada

1.0 We, mayors and municipal leaders from around the world meeting at the Fourth Municipal Leaders Summit on Climate Change submit a statement of solidarity as stewards of the Earth and agree that:

- Climate change is a major global challenge requiring urgent and concerted action and collaboration by all orders of government; and that,

- Climate change discussions, negotiations and actions are best informed by scientific evidence such as that provided by the Intergovernmental Panel on Climate Change (IPCC) with a particular focus on vulnerable continents and populations; and that,

- Municipal leaders have the extraordinary ability to change the current trend of global warming; and that,

- If substantial cooperation is exercised among all orders of government the resulting actions can be leveraged to realize the deep reductions needed to move toward climate stabilization.

2.0 We, mayors and municipal leaders, recognize that:

2.1 Local governments play a critical role to effectively reduce human induced greenhouse gas emissions knowing that the sustainable CO_2 emission rate for humankind is 0.5 tonnes eCO_2 per capita annually based on six billion inhabitants (IPCC).

2.2 Sustainable development and climate change are interdependent as articulated in the UN Millennium Development Goals.

2.3 Local policies and actions will meet or exceed targets set by sub-national and national governments to effect deep reductions and lead other sectors to execute the same.

2.4 Climate change impacts like floods, drought, water availability and quality, extreme heat, air pollution and infectious disease pose grave danger to public health and many local governments are already experiencing these effects.

2.5 The linkage between urban and rural communities driven by current development patterns offers opportunities to pursue poverty alleviation and mitigate inequitable impacts affected by climate change.

2.6 The buying power of local governments can accelerate the application and accessibility of clean technologies in the marketplace including renewable energy options.

2.7 The planet is warming. More severe and extreme weather events necessitate urgent action to ensure adequate mitigation and adaptation measures be taken to protect public health, strengthen infrastructure, apply appropriate urban and regional development plans, and advance economic development.

3.0 We, mayors and municipal leaders, commit to the following actions:

3.1 Implementation of policies and operational changes that, acknowledging the differential access to resources between cities in developed and developing countries, will achieve the emission reduction targets set forth in the International Youth Declaration of 30% by 2020 and 80% by 2050 based on 1990 levels, building upon the actions already taken by local governments that committed to a 20% reduction by 2010.

3.2 Establishing a system of accountability on these actions by reporting to the Conference of the Parties and Meeting of the Parties annually through 2012 detailing progress towards the targets.

3.3 Using uniform mechanisms to measure reductions for comparative analysis and verification.

3.4 Improving and advancing the exchange of data monitoring, skills, technologies, methods, tools, public education, and experiences to achieve emissions reductions, with specific reference to developing countries.

3.5 Minimization of the dependence on fossil fuel energy through shifting to sustainable land-use that:

- encourages public transit;
- diminishes the reliance on vehicular transport and single occupancy vehicles;
- improves energy efficiency.

3.6 Advancing partnerships and collaboration with national and sub-national governments, nongovernmental organizations, corporate and industrial sectors, as well as nongovernmental organizations and community groups, in order to multiply reduction potential.

4.0 We, mayors and municipal leaders, request that:

4.1 Local governments be recognized by the Conference of the Parties for the actions they have implemented and are continuing, tangibly to reduce greenhouse gas emissions. To this end, we request from the UNFCCC an allocation be granted to all Major Groups to strengthen and enhance an annual input process specific to local governments prior to future COP/MOP meetings.

4.2 National and sub-national governments: recognize the fundamental role of local governments in mitigating and adapting to climate change; partner with them to enhance their technical, human and financial capacity and legislative authority; and fully engage them when making strategic decisions on climate change policies.

4.3 Global trade regimes, credits and banking reserve rules be reformed to advance debt relief and incentives to implement polices and practices that reduce and mitigate climate change.

4.4 All national and sub-national governments commit to a process to negotiate an international climate change regime with deep reductions in greenhouse gas emissions enacted by 2012.

4.5 National and sub-national governments ensure that local governments have the opportunity to participate in emissions trading in accordance with evolving domestic and international trading systems.

For more information:

ICLEI—Local Governments for Sustainability
Montréal Summit Secretariat
City Hall, West Tower, 16th Floor
100 Queen Street West
Toronto, Ontario
M5H 2N2
Canada

Tel. +1-416/392-1390

Fax +1-416/392-1478

montreal.summit@iclei.org

www.iclei.org/montrealsummit

Appendix C

The U.S. Mayors Climate Protection Agreement

Mayors Endorsing the U.S. Mayors Climate Protection Agreement

State	City	Mayor Signing Agreement	Population
	Washington, D.C.	Anthony A. Williams	553,523
Alabama	Bessemer	Edward E. May	29,672
	Huntsville	Loretta Spencer	158,216
Alaska	Anchorage	Mark Begich	260,283
	North Pole	Jeffrey James Jacobson	1,778
	Shishmaref	Stanley Tocktoo	562
Arizona	Flagstaff	Joseph C. Donaldson	55,173
	Tucson	Robert E. Walkup	486,699
Arkansas	Fayetteville	Dan Coody	58,047
	Little Rock	Jim Dailey	183,133
	North Little Rock	Patrick Henry Hays	60,433
California	Albany	Allan Maris	16,444
	Aliso Viejo	Karl P. Warkowmski	45,000
	Arcata	Michael Machi	16,651
	Atascadero	Tom O'Malley	27,158

(continued)

State	City	Mayor Signing Agreement	Population
California	Atherton	Charles E. Marsala	7,177
	Berkeley	Tom Bates	102,743
	Burbank	Jef Vander Borght	100,316
	Capitola	Bruce R. Arthur	10,204
	Cerritos	Janet Abelson	51,488
	Chico	Scott Gruendl	59,954
	Chino	Dennis R. Yates	67,168
	Chula Vista	Stephen C. Padilla	203,000
	Cloverdale	Gail Pardini-Plass	7,275
	Cotati	Lisa Moore	6,400
	Cupertino	Richard Lowenthal	50,546
	Del Mar	Jerry Finnell	4,442
	Dublin	Janet Lockhart	34,500
	Fremont	Robert "Bob" Wasserman	203,413
	Hayward	Roberta Cooper	140,030
	Healdsburg	Jason Liles	11,101
	Hemet	Roger Meadows	58,812
	Hermosa Beach	Sam Y. Edgerton III	19,500
	Irvine	Beth Krom	143,072
	Lakewood	Joseph Esquivel	79,345
	Long Beach	Beverly O'Neill	471,000
	Los Altos Hills	Breene Kerr	8,122
	Los Angeles	Antonio Villaraigosa	3,845,541

State	City	Mayor Signing Agreement	Population
	Mill Valley	Anne B. Solem	13,286
	Moorpark	Patrick Hunter	35,894
	Monterey Park	Mike Eng	60,051
	Morgan Hill	Dennis Kennedy	33,556
	Morro Bay	Janice Peters	10,350
	Novato	Bernard H. Meyers	47,630
	Oakland	Jerry Brown	399,484
	Pacific Grove	Daniel E. Cort	15,091
	Palo Alto	Judy Kleinberg	58,598
	Pasadena	Bill Bogaard	143,731
	Petaluma	David Glass	54,548
	Pleasanton	Jennifer Hosterman	63,654
	Portola Valley	B. Stephen Toben	4,417
	Richmond	Irma L. Anderson	99,216
	Rohnert Park	Jake Mackenzie	42,236
	Sacramento	Heather Fargo	407,018
	San Bruno	Larry Franzella	40,165
	San Diego	Jerry Sanders	1,223,400
	San Francisco	Gavin Newsom	776,733
	San Jose	Ron Gonzales	904,522
	San Leandro	Shelia Young	79,452
	San Luis Obispo	Dave F. Romero	44,174
	San Mateo	Jan Epstein	92,482

(continued)

133

State	City	Mayor Signing Agreement	Population
California	San Rafael	Albert J. Boro	56,063
	Santa Ana	Miguel A. Pulido	337,977
	Santa Barbara	Marty Blum	92,325
	Santa Cruz	Mike Rotkin	54,593
	Santa Monica	Pam O'Connor	84,084
	Santa Rosa	Jane Bender	147,595
	Sausalito	Ronald P. Albert	7,228
	Sebastopol	Larry Robinson	7,774
	Sonoma	Larry Barnet	9,354
	Stockton	Edward J. Chavez	243,771
	Thousand Oaks	Claudia Bill-de la Peña	117,005
	Vallejo	Anthony J. Intintoli, Jr.	116,760
	West Hollywood	Abbe Land	35,716
	West Sacramento	Christopher Cabaldon	32,250
	Windsor	Steve Allen	24,180
Colorado	Aspen	Helen Kalin Klanderud	5,914
	Basalt	Leroy Duroux	3,007
	Boulder	Mark Ruzzin	94,673
	Denver	John W. Hickenlooper	554,636
	Durango	Sidny Zink	15,501
	Frisco	Bernie Zurbriggen	2,433
	Glenwood Springs	Bruce Christensen	8,564
	Gunnison	Stu Ferguson	5,319
	Telluride	John Pryor	2,221

State	City	Mayor Signing Agreement	Population
Connecticut	Bridgeport	John M. Fabrizi	139,529
	Easton	Wiliam J. Kupinse, Jr.	7,272
	Fairfield	Kenneth A. Flatto	57,340
	Hamden	Carl Amento	56,913
	Hartford	Eddie A. Perez	121,578
	Mansfield	Elizabeth C. Paterson	24,228
	Middletown	Domenique S. Thornton	43,167
	New Haven	John DeStefano, Jr.	123,626
	Stamford	Dannel P. Malloy	117,083
	Stratford	James R. Miron	49,976
Delaware	Wilmington	James M. Baker	72,664
Florida	Delray Beach	Jeff Perlman	60,020
	Gainesville	Pegeen Hanrahan	95,447
	Hallandale Beach	Joy Cooper	34,282
	Holly Hill	William D. Arthur	12,119
	Hollywood	Mara Giulianti	139,357
	Key Biscayne	Robert Oldakowski	10,507
	Key West	Jimmy Weekley	25,478
	Lauderhill	Richard J. Kaplan	57,585
	Miami	Manuel A. Diaz	362,470
	Miramar	Lori C. Moseley	72,739
	North Miami	Kevin Burns	59,880
	Oakland Park	Steven R. Arnst	30,966

(*continued*)

State	City	Mayor Signing Agreement	Population
Florida	Parkland	Michael Udine	22,145
	Pembroke Pines	Frank C. Ortis	137,427
	Pompano Beach	John C. Rayson	78,191
	Port St. Lucie	Robert E. Minsky	88,769
	Sunrise	Steven B. Feren	85,779
	Tallahassee	John Marks	150,624
	Tamarac	Joe Schreiber	55,588
	West Palm Beach	Lois J. Frankel	82,103
Georgia	Athens	Heidi Davison	101,489
	Atlanta	Shirley Franklin	416,474
	East Point	Patsy Jo Hilliard	39,595
	Macon	C. Jack Ellis	97,255
	Tybee Island	Jason Buelterman	3,400
Hawaii	Hilo	Harry Kim	135,499
	Honolulu	Mufi Hannemann	423,475
	Kauai	Bryan J. Bapitste	48,000
	Maui	Alan M. Arakawa	97,100
Idaho	Boise	David Bieter	185,787
Illinois	Carol Stream	Ross Ferraro	40,438
	Chicago	Richard M. Daley	2,862,244
	Highland Park	Michael D. Belsky	31,365
	Rock Island	Mark W. Schwiebert	39,684
	Schaumburg	Al Larson	75,386
	Waukegan	Richard H. Hyde	87,901

State	City	Mayor Signing Agreement	Population
Indiana	Bloomington	Mark Kruzan	69,291
	Carmel	James Brainard	37,733
	Columbus	Fred L. Armstrong	39,059
	Fort Wayne	Graham A. Richard	205,727
	Gary	Scott L. King	102,746
	Michigan City	Chuck Oberlie	32,900
Iowa	Des Moines	T. M. Franklin Cownie	198,682
	Dubuque	Roy D. Buol	57,686
	Sioux City	Craig S. Berenstein	85,013
Kansas	Lawrence	Dennis Highberger	81,000
	Topeka	James A. McClinton	122,377
Kentucky	Lexington	Teresa Isaac	260,512
	Louisville Metro	Jerry E. Abramson	694,000
Louisiana	Alexandria	Edward G. Randolph, Jr.	46,342
	New Orleans	C. Ray Nagin	484,674
Maine	Belfast	Michael D. Hurley	6,500
	Biddeford	Wallace H. Nutting	22,072
	Portland	James I. Cohen	64,249
	Saco	Mark D. Johnston	18,230
Maryland	Annapolis	Ellen O. Moyer	35,838
	Baltimore	Martin O'Malley	651,154
	Chevy Chase	William H. Hudnut	31,000
	Gaithersburg	Sidney A. Katz	52,613

(continued)

State	City	Mayor Signing Agreement	Population
Maryland	Rockville	Larry Giammo	57,402
	Sykesville	Jonathan Herman	4,197
Massachusetts	Boston	Thomas M. Menino	589,141
	Cambridge	Michael A. Sullivan	101,355
	Gloucester	John Bell	30,273
	Hull	John D. Reilly, Jr.	11,320
	Malden	Richard C. Howard	56,340
	Medford	Michael J. McGlynn	55,765
	Melrose	Robert J. Dolan	26,365
	Newton	David B. Cohen	83,829
	New Bedford	Scott W. Lang	93,768
	Pitssfield	James M. Ruberto	45,793
	Provincetown	Keith A. Bergman	3426
	Somerville	Joseph A. Curtatone	77,478
	Truro	Alfred Gaechter	2,146
	Worcester	Timothy P. Murray	172,648
Michigan	Ann Arbor	John Hieftje	114,024
	Berkley	Marilyn Stephen	15,236
	Ferndale	Robert Porter	22,100
	Grand Rapids	George Heartwell	197,800
	Marquette	Tony Tollefson	20,581
	Southfield	Brenda L. Lawrence	78,296
Minnesota	Apple Valley	Mary Hamann-Roland	45,527
	Burnsville	Elizabeth B. Kautz	60,220

State	City	Mayor Signing Agreement	Population
	Duluth	Herb Bergson	86,918
	Eden Prairie	Nancy Tyra-Lukens	54,901
	Milan	Ron Anderson	323
	Minneapolis	R. T. Rybak	382,618
	Rochester	Ardell F. Brede	85,806
	St. Paul	Chris Coleman	287,151
	Turtle River	Gary A. Burger	79
Mississippi	Meridian	John Robert Smith	39,968
Missouri	Clayton	Ben Uchitelle	15,944
	Florissant	Robert G. Lowery, Sr.	50,497
	Kansas City	Kay Barnes	441,545
	Kirkwood	Mike Swoboda	27,324
	Maplewood	Mark Langston	9,600
	St. Louis	Francis G. Slay	348,189
	Sunset Hills	James A. Hobbs	8,267
	University City	Joseph L. Adams	37,428
Montana	Billings	Charles F. Tooley	89,847
	Bozeman	Jeff Krauss	33,535
	Missoula	Mike Kadas	57,053
Nebraska	Bellevue	Jerry Ryan	44,382
	Lincoln	Coleen J. Seng	225,581
	Omaha	Mike Fahey	390,007

State	City	Mayor Signing Agreement	Population
Nevada	Henderson	James B. Gibson	175,381
	Las Vegas	Oscar B. Goodman	478,434
	Reno	Robert Cashell	180,480
	Sparks	Geno R. Martini	66,346
New Hampshire	Dover	J. Michael Joyal	28,486
	Hanover	Brian F. Welsh	11,156
	Keene	Michael E. J. Blastos	22,563
	Manchester	Frank C. Guinta	107,006
	Nashua	Bernard A. Streeter	86,605
New Jersey	Bayonne	Joseph V. Doria, Jr.	61,842
	Bloomfield	Raymond J. McCarthy	47,683
	Brick	Joseph C. Scarpelli	76,119
	Cliffside Park Borough	Gerald A. Calabrese	23,035
	Closter	Fred Pitofsky	8,400
	Cranford	Daniel J. Aschenbach	22,478
	Elizabeth	J. Christian Bollwage	120,568
	Ewing	Wendell E. Pribila	35,707
	Galloway Township	Thomas Bassford	35,833
	Hamilton	Glen D. Gilmore	87,109
	Hightstown	Robert F. Patten	5,326
	Hope	Timothy C. McDonough	1,891
	Hopewell Borough	David R. Nettles	2,051
	Hopewell Township	Vanessa Sandom	17,582

State	City	Mayor Signing Agreement	Population
	Kearny	Alberto G. Santos	40,513
	Linwood	Richard L. DePamphilis III	7,398
	Long Hill Township	Gina Genovese	8,787
	Maple Shade	John D. Galloway	19,502
	Montclair	Edward A. Remsen	38,977
	Newark	Sharpe James	273,546
	Ocean City	Salvatore Perillo	15,330
	Plainfield	Albert T. McWilliams	47,829
	Ringwood	Joanne Atlas	12,800
	Robbinsville	Dave Fried	11,000
	Teaneck	Elie Y. Katz	39,260
	Township of Elk	William J. Rainey, Jr.	3,792
	West Orange	John F. McKeon	44,943
	Westfield	Gregory S. McDermott	29,644
	West Milford	Joseph A. DiDonato	28,181
	West Windsor	Shing-Fu Hsueh	24,458
New Mexico	Alamogordo	Donald E. Carroll	35,582
	Albuquerque	Martin J. Chavez	448,607
	Capitan	Sam Hammons	1,486
	Las Cruces	William M. Mattiace	74,267
	Ruidoso	L. Ray Nunley	8,812
	Santa Fe	Larry A. Delgado	62,203

State	City	Mayor Signing Agreement	Population
New York	Albany	Gerald D. Jennings	95,658
	Babylon	Steve Bellone	216,230
	Buffalo	Anthony M. Masiello	292,648
	Greenburgh	Paul Feiner	89,514
	Hempstead	Wayne Hall	56,554
	Hudson	Richard F. Tracy	7,145
	Irvington, Village of	Dennis P. Flood	6,650
	Ithaca	Carolyn Peterson	29,287
	Mount Vernon	Ernest D. Davis	68,381
	New York	Michael R. Bloomberg	8,104,079
	Niagara Falls	Vince V. Anello	55,593
	Rochester	William A. Johnson, Jr.	219,773
	Rockville Centre	Eugene J. Murray	24,568
	Schenectady	Brian Stratton	61,821
	Syracuse	Matthew J. Driscoll	147,306
	White Plains	Joseph M. Delfino	53,077
North Carolina	Asheville	Charles R. Worley	68,889
	Boone	Loretta Clawson	13,192
	Carrboro	Mark Chilton	16,800
	Chapel Hill	Kevin C. Foy	48,715
	Durham	William V. "Bill" Bell	187,035
	Highlands	Dr. Don Mullen	941
	Wilmington	Bill Saffo	75,838

State	City	Mayor Signing Agreement	Population
North Dakota	Fargo	Dennis R. Walaker	90,599
Ohio	Akron	Donald L. Plusquellic	217,074
	Brooklyn	Kenneth E. Patton	11,586
	Cincinatti	Mark Mallory	331,285
	Cleveland	Frank G. Jackson	478,403
	Cleveland Heights	Edward J. Kelley	49,958
	Dayton	Rhine L. McLin	166,179
	Garfield Heights	Thomas J. Longo	30,734
	North Olmsted	Thomas O'Grady	34,113
	South Euclid	Georgine Welo	22,210
	Toledo	Jack Ford	313,619
Oklahoma	Norman	Harold Haralson	95,694
Oregon	Ashland	John W. Morrison	20,755
	Beaverton	Rob Drake	76,129
	Corvallis	Helen Berg	49,322
	Eugene	Kitty Piercy	137,893
	Lake Oswego	Judie Hammerstad	35,278
	Lincoln City	Lori Hollingsworth	7,849
	Portland	Tom Potter	529,121
Pennsylvania	Allentown	Ed Pawlowski	106,632
	Bethlehem	John B. Callahan	71,329
	Easton	Philip B. Mitman	26,267
	Erie	Richard E. Filippi, Esq.	103,717

(continued)

State	City	Mayor Signing Agreement	Population
Pennsylvania	Philadelphia	John F. Street	1,470,151
Rhode Island	Pawtucket	James E. Doyle	72,958
	Providence	David N. Cicilline	173,618
	Warwick	Scott Avedisian	85,808
South Carolina	Charleston	Joseph P. Riley, Jr.	96,650
	Columbia	Robert D. Coble	116,278
	Greenville	Knox H. White	56,002
	Sumter	Joseph T. McElveen, Jr.	39,643
Tennessee	Chattanooga	Ron Littlefield	155,554
	Cookeville	Charles Womack	27,648
	Nashville	Bill Purcell	592,099
Texas	Arlington	Robert Cluck	332,969
	Austin	Will Wynn	681,804
	Dallas	Laura Miller	1,188,580
	Denton	Euline Brock	80,537
	Euless	Mary Lib Saleh	46,005
	Frisco	E. Michael Simpson	33,714
	Hurst	Richard Ward	36,273
	Laredo	Elizabeth G. "Betty" Flores	176,576
	McKinney	Bill Whitfield	54,369
	Richardson	Gary A. Slagel	91,802
	Sugar Land	David G. Wallace	63,328
Utah	Moab	David Sakrison	4,825

State	City	Mayor Signing Agreement	Population
	Park City	Dana Williams	7,300
	Salt Lake City	Ross "Rocky" C. Anderson	181,743
Vermont	Burlington	Peter A. Clavelle	38,889
Virginia	Alexandria	William D. "Bill" Euille	128,283
	Blacksburg	Ron Rordam	39,573
	Charlottesville	David E. Brown	45,049
	Richmond	L. Douglas Wilder	197,790
	Virginia Beach	Meyera E. Oberndorf	425,257
	Williamsburg	Jeanne Zeidler	11,751
Washington	Auburn	Peter B. Lewis	40,314
	Bainbridge Island	Darlene Kordonowy	20,300
	Battle Ground	John G. Idsinga	13,237
	Bellingham	Mark Asmundson	67,171
	Burien	Noel Gibb	31,881
	Edmonds	Gary Haakenson	39,515
	Everett	Ray Stephanson	91,488
	Issaquah	Ava Frisinger	11,212
	Kirkland	Mary-Alyce Burleigh	45,054
	Lacey	Virgil Clarkson	31,226
	Lake Forest Park	David Hutchinson	12,474
	Lynnwood	Mike McKinnon	33,847
	Olympia	Mark Foutch	42,514

(continued)

State	City	Mayor Signing Agreement	Population
Washington	Redmond	Rosemarie M. Ives	45,256
	Renton	Kathy Keolker-Wheeler	53,840
	Sammamish	Michele Petitti	34,364
	Seattle	Greg Nickels	571,480
	Shoreline	Robert L. Ransom	53,296
	Tacoma	Bill Baarsma	193,556
	Vancouver	Royce E. Pollard	143,560
West Virginia	Shepherdstown	Lance Dom	1,158
Wisconsin	Ashland	Friedrich P. Schnook	8,795
	Greenfield	Michael Neitzke	36,059
	La Crosse	John D. Medinger	51,818
	Madison	Dave Cieslewicz	208,054
	Milwaukee	Tom Barrett	596,974
	New Berlin	Jack F. Chiovatero	38,220
	Racine	Gary Becker	81,855
	River Falls	Don Richards	13,019
	Stevens Point	Gary Wescott	24,298
	Washburn	Friedrich P. Schnook	2,298
	Wauwatosa	Theresa M. Estness	47,271
	West Allis	Jeannette Bell	61,254
Wyoming	Jackson	Mark Barron	9,038
		Total:	**55,020,847**

SOURCE: Information from the City of Seattle, current as of January 10,
2007, www.seattle.gov/mayor/climate/quotes.htm#mayors
(accessed January 14, 2007).

Endorsing the U.S. Mayors Climate Protection Agreement: U.S. Conference of Mayors 2005 Adopted Resolutions

WHEREAS, the U.S. Conference of Mayors has previously adopted strong policy resolutions calling for cities, communities and the federal government to take actions to reduce global warming pollution; and

WHEREAS, the Inter-Governmental Panel on Climate Change (IPCC), the international community's most respected assemblage of scientists, has found that climate disruption is a reality and that human activities are largely responsible for increasing concentrations of global warming pollution; and

WHEREAS, recent, well-documented impacts of climate disruption include average global sea level increases of four to eight inches during the 20th century; a 40 percent decline in Arctic sea-ice thickness; and nine of the ten hottest years on record occurring in the past decade; and

WHEREAS, climate disruption of the magnitude now predicted by the scientific community will cause extremely costly disruption of human and natural systems throughout the world including: increased risk of floods or droughts; sea-level rises that interact with coastal storms to erode beaches, inundate land, and damage structures; more frequent and extreme heat waves; more frequent and greater concentrations of smog; and

WHEREAS, on February 16, 2005, the Kyoto Protocol, an international agreement to address climate disruption, went into effect in the 141 countries that have ratified it to date; 38 of those countries are now legally required to reduce greenhouse gas emissions on average 5.2 percent below 1990 levels by 2012; and

WHEREAS, the United States of America, with less than 5 percent of the world's population, is responsible for producing approximately 25 percent of the world's global warming pollutants; and

WHEREAS, the Kyoto Protocol emissions reduction target for the United States would have been 7 percent below 1990 levels by 2012; and

WHEREAS, many leading U.S. companies that have adopted greenhouse gas reduction programs to demonstrate corporate social responsibility have also publicly expressed preference for the United States to adopt precise and mandatory emissions targets and timetables as a means by which to remain competitive in the international marketplace, to mitigate financial risk, and to promote sound investment decisions; and

WHEREAS, state and local governments throughout the United States are adopting emission reduction targets and programs and that this leadership is bipartisan, coming from Republican and Democratic governors and mayors alike; and

WHEREAS, many cities throughout the nation, both large and small, are reducing global warming pollutants through programs that provide economic and quality of life benefits such as reduced energy bills, green space preservation, air quality improvements, reduced traffic congestion, improved transportation choices, and economic development and job creation through energy conservation and new energy technologies; and

WHEREAS, mayors from around the nation have signed the U.S. Mayors Climate Protection Agreement which, as amended at the 73rd Annual U.S. Conference of Mayors meeting, reads:

The U.S. Mayors Climate Protection Agreement

A. We urge the federal government and state governments to enact policies and programs to meet or beat the target of reducing global warming pollution levels to 7 percent below 1990 levels by 2012, including efforts to: reduce the United States' dependence on fossil fuels and accelerate the development of clean, economical energy resources and fuel-efficient technologies such as conservation, methane recovery for energy generation, waste to energy, wind and solar energy, fuel cells, efficient motor vehicles, and biofuels;

B. We urge the U.S. Congress to pass bipartisan greenhouse gas reduction legislation that includes 1) clear timetables and emissions limits and 2) a flexible, market-based system of tradable allowances among emitting industries; and

C. We will strive to meet or exceed Kyoto Protocol targets for reducing global warming pollution by taking actions in our own operations and communities such as:

1. Inventory global warming emissions in city operations and in the community, set reduction targets, and create an action plan.

2. Adopt and enforce land-use policies that reduce sprawl, preserve open space, and create compact, walkable urban communities;

3. Promote transportation options such as bicycle trails, commute trip reduction programs, incentives for car pooling, and public transit;

4. Increase the use of clean, alternative energy by, for example, investing in "green tags," advocating for the development of renewable energy resources, recovering landfill methane for energy production, and supporting the use of waste to energy technology;

5. Make energy efficiency a priority through building code improvements, retrofitting city facilities with energy efficient lighting, and urging employees to conserve energy and save money;

6. Purchase only Energy Star equipment and appliances for city use;

7. Practice and promote sustainable building practices using the U.S. Green Building Council's LEED program or a similar system;

8. Increase the average fuel efficiency of municipal fleet vehicles; reduce the number of vehicles; launch an employee education program including anti-idling messages; convert diesel vehicles to bio-diesel;

9. Evaluate opportunities to increase pump efficiency in water and wastewater systems; recover wastewater treatment methane for energy production;

10. Increase recycling rates in city operations and in the community;

11. Maintain healthy urban forests; promote tree planting to increase shading and to absorb CO_2; and

12. Help educate the public, schools, other jurisdictions, profes-
    sional associations, business and industry about reducing
    global warming pollution.

NOW, THEREFORE, BE IT RESOLVED that The U.S. Conference of Mayors endorses the U.S. Mayors Climate Protection
Agreement as amended by the 73rd annual U.S. Conference of
Mayors meeting and urges mayors from around the nation to
join this effort.

BE IT FURTHER RESOLVED, The U.S. Conference of Mayors will work in conjunction with ICLEI Local Governments for
Sustainability and other appropriate organizations to track
progress and implementation of the U.S. Mayors Climate Protection Agreement as amended by the 73rd annual U.S. Conference of Mayors meeting.

U.S. Jurisdictions Participating in ICLEI's Cities for Climate Protection Campaign, February 1, 2007

Alachua County, Florida

Alameda, California

Alameda County, California

Albany, California

Albuquerque, New Mexico

Amherst, Massachusetts

Anchorage, Alaska

Ann Arbor, Michigan

Annapolis, Maryland

Arcata, California

Arlington County, Virginia

Arlington, Massachusetts

Arlington, Texas

Asheville, North Carolina

Ashland, Oregon

Aspen, Colorado

Atlanta, Georgia

Augusta, Maine

Austin, Texas

Babylon, New York

Barnstable, Massachusetts

Bellingham, Washington

Belmar, New Jersey

Belmont, Massachusetts

Berkeley, California

Blacksburg, Virginia

Boise, Idaho

Boston, Massachusetts

Boulder, Colorado

Brattleboro, Vermont

Bridgeport, Connecticut

Brookline, Massachusetts

Broward County, Florida

Buffalo, New York

Burien, Washington

Burlington, Vermont

Cambridge, Massachusetts

Carbondale, Colorado

Carol Stream, Illinois

Carrboro, North Carolina

CCRPA, Connecticut

Central Massachusetts Planning, Massachusetts

Chapel Hill, North Carolina

Charleston, South Carolina

Chattanooga, Tennessee

Chevy Chase, Maryland

Chicago, Illinois

Chittenden County, Vermont

Chula Vista, California

Cloverdale, California

College Park, Maryland

Columbia, South Carolina

Cooperstown, New York

Corvallis, Oregon

Costra County, California

Cotati, California

Dallas, Texas

Dane County, Wisconsin

Davis, California

Decatur, Georgia

Delta County, Michigan

Denton, Texas

Denver, Colorado

Des Moines, Iowa

Devens, Massachusetts

Duluth, Minnesota

Durham, North Carolina

Edmonds, Washington

El Cerrito, California

Emeryville, California

Epping, New Hampshire

Eugene, Oregon

Fairfax, California

Fairfield, Connecticut

Falmouth, Massachusetts

Farmington, Maine

Fayetteville, Arkansas

Fort Bragg, California

Fort Collins, Colorado

Fort Wayne, Indiana

Gainesville, Florida

Georgetown, South Carolina

Gloucester, Massachusetts

Golden, Colorado

Grand Rapids, Michigan

Greenburgh, New York

Gunnison County, Colorado

Hamden, Connecticut

Hamilton, New Jersey

Harrisonburg, Virginia

Hartford, Connecticut

Healdsburg, California

Hennepin County, Minnesota

Hillsborough County, Florida

Honolulu, Hawaii

Houston, Texas

Hull, Massachusetts

Huntington, New York

Ipswich, Massachusetts

Irvine, California

Ithaca, New York

Jackson, Wyoming

Kansas City, Missouri

Keene, New Hampshire

King County, Washington

Kirkland, Washington

LaConner, Washington

Langley, Washington

Lenox, Massachusetts

Little Rock, Arkansas

Los Angeles, California

Louisville Metro, Kentucky

Lynn, Massachusetts

Madison, Wisconsin

Maplewood, New Jersey

Marin County, California

Marin Municipal Water District, California

Medford, Massachusetts

Memphis, Tennessee

Mendocino County, California

Mesa, Arizona

Miami Beach, Florida

Miami-Dade County, Florida

Middlebury, Vermont

Milwaukee, Wisconsin

Minneapolis, Minnesota

Missoula, Montana

Montgomery County, Maryland

Montpelier, Vermont

Mount Rainier, Maryland

Mount Vernon, New York

Multnomah County, Oregon

Muncie, Indiana

Nashua, New Hampshire

Natick, Massachusetts

New Britain, Connecticut

New Haven, Connecticut

New Orleans, Louisiana

New Paltz, New York

New Rochelle, New York

New York, New York

Newark, California

Newark, New Jersey

Newburyport, Massachusetts

Newton, Massachusetts

Northampton, Massachusetts

Northfield, Minnesota

Novato, California

Oakland, California

Oak Harbor, Washington

Olympia, Washington

Oneonta, New York

Orange County, Florida

Orange County, North Carolina

Overland Park, Kansas

Palo Alto, California

Pawtucket, Rhode Island

Petaluma, California

Philadelphia, Pennsylvania

Piedmont, California

Pioneer Valley Planning, Massachusetts

Pittsburgh, Pennsylvania

Pittsfield, Massachusetts

Plainville, Connecticut

Plano, Texas

Point Arena, California

Portland, Maine

Portland, Oregon

Portola Valley, California

Prince George's County, Maryland

Providence, Rhode Island

Provincetown, Massachusetts

Ramsey County, Minnesota

Reading, Massachusetts

Riviera Beach, Florida

Roanoke, Virginia

Rohnert Park, California

Sacramento, California

Saint Paul, Minnesota

Salem, Massachusetts

Salt Lake City, Utah

Salt Lake County, Utah

San Anselmo, California

San Antonio, Texas

San Diego, California

San Francisco, California

San Jose, California

San Leandro, California

San Miguel County, Colorado

San Rafael, California

Santa Barbara, California

Santa Clara County, California

Santa Cruz, California

Santa Fe, New Mexico

Santa Monica, California

Santa Rosa, California

Sarasota County, Florida

Saratoga Springs, New York

Sausalito, California

Schenectady County, New York

Seattle, Washington

Sebastopol, California

Shutesbury, Massachusetts

Somerville, Massachusetts

Sonoma City, California

Sonoma County, California

Spokane County, Washington

Spokane, Washington

Springfield, Massachusetts

Stamford, Connecticut

Suffolk County, New York

Syracuse, New York

Tacoma, Washington

Takoma Park, Maryland

Tampa, Florida

Toledo, Ohio

Tompkins County, New York

Tucson, Arizona

Tumwater, Washington

Union City, California

Washtenaw County, Michigan

Watertown, Massachusetts

West Chester, Pennsylvania

West Hollywood, California

Westchester County,
New York

Weston, Connecticut

Whatcom County, Washington

Williamstown, Massachusetts

Willits, California

Winchester, Massachusetts

Windham, Connecticut

Windsor, California

Windsor, Connecticut

Worcester, Massachusetts

Appendix E

ICLEI's Quantification Methodology

ICLEI—Local Governments for Sustainability has developed an emissions quantification protocol for local governments participating in the Cities for Climate Protection (CCP) Campaign. This protocol provides jurisdictions with an easily implemented set of guidelines to assist them in the quantification of the greenhouse gas emissions and reductions from their internal operations and from their jurisdictions as a whole. By developing common conventions and a standardized approach, ICLEI seeks to make it easier for campaign participants to fulfill their commitments to the five CCP milestones and in doing so achieve tangible reductions in greenhouse gas emissions. The standardized approach described in this protocol facilitates comparisons among local governments and the aggregation and reporting of results being achieved by Local Action Plans of diverse communities.

ICLEI's protocol quantifies the emissions and reductions of the most common greenhouse gases (CO_2, CH_4, and N_2O) and criteria air pollutants (CO, SO_x, NO_x, PM_{10}, and VOC). The calculations are run both for the community as a whole and for the local government's internal operations. Within the community, consideration is given to the residential, commercial, and industrial, transportation, and waste sectors. From the government's internal operations, consideration is given to the buildings, vehicle fleet, employee commute, waste, water/sewage, and signals/streetlight sectors.

Emissions levels are calculated by collecting data on all energy used, fuel burned, vehicle miles traveled, and/or waste generated. This input data is then multiplied by a number of emission factors to calculate the emissions level. These emission factors have been derived from a number of sources, including the U.S. Environmental Protection Agency, Department of Energy, the Intergovernmental Panel on Climate Change, and original research conducted by ICLEI and its consultants. The following formula is used to calculate the emissions level:

energy use (kWh electricity) $\times$ emissions factor (tons eCO_2 per kWH) = emissions (tons eCO_2)

Emissions reductions are computed in a similar fashion. First, the emissions in the "before" case are quantified using the methodology described above. Then, the emissions in the "after" case are calculated. The difference between the emissions in these two cases equals the change in emissions due to the measure put into place. For example, if a city planned to switch to LED traffic signals, the city would use the amount of energy used in a year as the "before" case. The amount of energy for the "after" case could be either measured directly after the lights had been changed out or calculated based on the percentage less energy the LED bulbs use than standard incandescent bulbs. The following formula is used to calculate the change in emissions:

[energy use before (kWh electricity) × emissions factor (tons eCO_2 per kWH) = emissions before (tons eCO_2)]

−

[energy use after (kWh green electricity) × emissions factor (tons eCO_2 per kWH) = emissions after (tons eCO_2)]

=

emissions reduction (tons eCO_2)

For more information on the details of these calculations, visit the ICLEI website, http://www.iclei.org/usa.

U.S. Green Building Council LEED Users in the Governmental Sectors

As mentioned in Chapter Five, numerous local governments, as well as state governments and federal programs, have adopted policies requiring new buildings to meet Leadership in Energy and Environmental Design (LEED) guidelines. Below is a list of federal, state, and local initiatives regarding LEED, as of August 2006.

For updates, contact 202-828-7422 or visit www.usgbc.org.

Legislation (6)

Arkansas, State of

Baltimore County, Maryland

Maryland, State of

Nevada, State of

New York City

Washington State

Ordinance (20)

Alameda County, California

Atlanta, Georgia

Calabasas, California

Chapel Hill, North Carolina

Chatham County, Georgia

Cook County, Illinois

Cranford, New Jersey

Frisco, Texas

Gainesville, Florida

Honolulu, Hawaii

Los Angeles, California

New York, New York

Normal, Illinois

Oakland, California

Pasadena, California

Pleasanton, California

San Francisco, California

San Mateo, California

San Jose, California

Santa Monica, California

Resolution (16)

Austin, Texas

Berkeley, California

Bowie, Maryland

Chicago, Illinois

Dallas, Texas

Eugene, Oregon

Houston, Texas

Kansas City, Missouri

Portland, Oregon

Sacramento, California

Sarasota, Florida

Scottsdale, Arizona

Sufford County, New York

Tucson, Arizona

Tybee Island, Georgia

Vancouver, British Columbia

Executive Order (14)

Albuquerque, New Mexico

Arizona

California

Colorado

Maine

Maryland

Michigan

New Jersey

New Mexico

New York (does not require certification)

King County, Washington

Rhode Island

Salt Lake City, Utah

Wisconsin

Incentives for LEED (17)

Tax

Baltimore County, Maryland

Chatham County, Georgia

Maryland

Nevada

New York

Oregon

Portland, Oregon

Pasadena, California

Density Bonus

Acton, Massachusetts

Arlington, Virginia

Expedited Permit Review

Gainesville, Florida

Issaquah, Washington

Santa Monica, California

Sarasota County, Florida

Grant Program

Pennsylvania—schools

Santa Monica, California

Other

Cranford, New Jersey (incentives vary by request)

Private Sector Initiatives (6)

Boulder, Colorado

Brisbane, California

Calabasas, California

Normal, Illinois

Pasadena, California

Pleasanton, California

Green Building Policies Including LEED (10)

Boston, Massachusetts

Calgary, Alberta

Grand Rapids, Michigan

Long Beach, California

Madison, Wisconsin

Phoenix, Arizona

Princeton, New Jersey

San Diego, California

Seattle, Washington

Washington, DC, Department of Parks and Recreation

Federal Initiatives (9)

Department of Agriculture—Forest Services

Department of Defense

- U.S. Air Force
- U.S. Army
- U.S. Navy

U.S. Department of Energy

U.S. Department of General Services

Department of the Interior

U.S. Department of State

U.S. Environmental Protection Agency

Emission Reduction Measures in Fort Collins Climate Action Plan

The following tables include the measures outlined in Fort Collins original climate action plan and the estimated greenhouse gas reductions they would achieve in the target year. Updates on which actions have been implemented and their progress in reducing emissions to date are reported in the city's biannual progress reports. These reports are available on line at http://www .fcgov.com/climateprotection/policy.php.

Existing Measures	Tons CO_2 Reduced in 2010
Vehicle Miles Traveled goal: not exceed population growth rates	337,676
Business recycling (apply 1998 per capita rate to 2010 population)	41,735
1997 City Energy Code (existing and projected benefits)	40,436
Curbside recycling (apply 1998 per capita rate to 2010 population)	39,732
Climate Wise for businesses	38,390
Methane flaring and heat recovery at city's water reclamation plant	35,607
Sequestration of CO_2 by all trees in Fort Collins	21,071
Electricity distribution system improvements	15,189
CSU utility system (energy conservation programs)— benefit from existing programs	12,524
CSU's Industrial Assessment Center (savings in 2010 from existing projects)	4,429

(continued)

Existing Measures	Tons CO_2 Reduced in 2010
Wind Power Pilot Program	4,013
Wind Phase II (2.5 more turbines)	5,128
Promote telecommuting	3,076
Poudre School District energy conservation programs (existing and projected benefits)	2,552
ZILCH with energy score (existing and projected benefits)	652
ZILCH without energy score (existing and projected benefits)	291
Lighting upgrades—city buildings: 1990–1998	257
Propane city fleet vehicles (assume 1998 use rates)	139
10% reduction of municipal solid waste in 2010	121
Natural areas shrub plantings	58
Converting to variable frequency drives (city government actions through 1998)	48
Consider accelerated TDM program: disincentives for drving	Unknown
ULEV and ZEV vehicles for city fleet including electric vehicles	Unknown
Clean Cities Program	Unknown
Pollution Prevention (P2) to promote energy efficiency in the commercial Sector	Unknown
Municipal pilot of environmentally preferable products	Unknown
"Green building" for the new city office building	Unknown
Existing Measures Total	**603,124**

Pending Measures	Tons CO_2 Reduced in 2010
50% solid waste diversion goal	112,787
Push for tighter national fuel efficiency (CAFE) standards	120,750
Second centralized recycling drop-off site	1,095
Fort Collins–Denver Commuter Rail	15–50,000
Expand Larimer County recycling center	18,834
Landfill gas to energy	84,307
Increase energy efficiency training for builders	20,840
Green-building program for residences	1,665
Trash districting	292
Parks satellite shop	13
Construction and demolition pilot	Unknown
Work with MAPO to encourage bulk purchasing	Unknown
Pending Measures Total	**373,154**

New Measures Type	Tons CO_2 Reduced in 2010
Replace traffic signals with LEDs	3,137
Continuation of wind program (5 more turbines)	10,255
Climate change education and outreach	40,553
Optimization of wastewater treatment motors/pumps	961
Reduce city government building energy use 15% below 1990 levels (per SqFt)	3,129
City Government purchase a portion of wind for own electricity needs (1 turbine)	2,051
Increase awareness of fuel consumption in city departments	62
Green building program for commercial construction	3,186
Push for mandatory renewable in electric deregulation (4% of all in 2010) or comparable energy conservation	71,561
Increase citywide tree planting (3,600 more trees)	125
Distribute bids and proposals electronically	3
Increase mortality age of city-owned trees	Unknown
New Measures Total	**135,023**

SOURCE: Information in these tables is from City of Fort Collins, Natural Resources Department, "City of Fort Collins Local Action Plan to Reduce Greenhouse Gas Emissions," 1999.

Portland Greenhouse Gas Reduction Measures

The following is a sample of the types of greenhouse gas reduction measures included in City of Portland/Multnomah County Global Warming Local Action Plan.

Focus Area	Goals and Target	Existing Measures	Proposed Measures
Policy research and education	To provide policy, research, and education to local agency staff and the community.	• Biannual report of greenhouse gas emissions to monitor change and adjust plan • Community outreach and education	• Inter-office communications and training sessions • Advocacy for national action on global warming
Energy efficiency	Increase energy efficiency to reduce energy use in facilities across all sectors by 10% to avoid 0.67 million metric tons.	• Block-by-block: free insulation for low-income single-family homes • Multi-family energy saving Program • BEST: Businesses for an Environmentally Sustainable Tomorrow	• Investment in efficiency measures with short-term paybacks • Public-private energy conservation partnerships • Green building standards for municipal construction, technical assistance,

(continued)

Focus Area	Goals and Target	Existing Measures	Proposed Measures
		offers assistance and awards	education, and financial incentives to builders • Convert traffic lights to LED bulbs • County-wide energy use best-practices performance and standards for equipment
Transportation, telecommunication, and access	Reduce per capita vehicle miles traveled by 10% below 1995 and improve average fuel efficiency of vehicles from 18.5 to 26 mpg to avoid 1.35 million metric tons	• Large employers required to reduce employee commute trips by 10% • Trip Reduction Incentives Program provides transit and carpool incentives to city employees	• Open three major light-rail lines • Develop more than 200 miles of bikeways • Launch the first modern streetcar line in the United States • Switch from diesel to biodiesel • Purchase hybrid vehicles for vehicle fleet
Renewable energy resources	Emphasize renewable energy resources to meet all growth in electricity load through renewable sources to avoid 0.54 million metric tons	• Capture methane from a wastewater treatment plant for neighboring industry and to heat buildings • City purchase renewable energy	• Renewable energy demonstration projects • Portland's construction of a fuel-cell electricity generator powered by methane from the city's

164

Focus Area	Goals and Target	Existing Measures	Proposed Measures
		• Tax credits for geothermal space- and water-heating systems	wastewater treatment plant to produce electricity to power 120 homes • Portland's purchase of 44 million kWh of wind power • Solar cells on maintenance vehicles to power tools without engine
Waste reduction and recycling	Promote waste reduction and recycling to minimize methane emissions from landfills and manufacturing processes to avoid 0.23 million metric tons	• Recover 60% of waste by 2005 • Increase residential curbside recycling • A commercial recycling ordinance (1996) requires every work site to set up and use a recycling system	• Require that 50% of solid waste from businesses be recycled • Sustainable Paper use policy to minimize paper generation from city's operations • Commercial food waste collection and recycling programs • Investigate and set standards to purchase recycled materials
Forestry and carbon offsets	Expand urban and rural forestry practices to avoid	• Urban Forestry Management Plan lists	• Support reforestation initiative by planting 3,000

(continued)

Focus Area	Goals and Target	Existing Measures	Proposed Measures
	0.31 million metric tons	strategies and recommended actions	acres of trees • Document benefits provided by urban forest cover and use data to inform policy decisions and get funding • Tree planting to maximize carbon offsets, energy conservation, air quality, stormwater management, and habitat benefits

SOURCE: Information in this table is a summary of some of the measures included in Portland Office of Sustainable Development, Multnomah County Department of Sustainable Community Development, "Local Action Plan on Global Warming: City of Portland & Multnomah County," 2001.

Notes

Introduction (pages ix–xii)

1. U.S. Environmental Protection Agency, statement of G. Tracy Mehan III before the Committee on Environment and Public Works, October 8, 2002, www .epa.gov/water/speeches/021008tm.html (accessed on October 17, 2006).

Chapter One: A Global Warming Overview (pages 1–13)

1. Robert Minsky, Mayor of Port St. Lucie, Florida, quoted in Seattle, www.seattle .gov/mayor/climate/quotes.htm (accessed on December 19, 2006).
2. Lester R. Brown, *The Earth Policy Reader* (New York: W.W. Norton and Company, 2002).
3. National Oceanic and Atmospheric Administration, "NOAA Reports 2006 Warmest Year on Record for U.S.: General Warming Trend, El Niño Contribute to Milder Winter Temps," www.noaanews.noaa.gov/stories2007/s2772.htm (accessed on February 3, 2007); National Aeronautics and Space Adminsitration, "2005 Warmest Year in over a Century," www.nasa.gov/vision/earth/environment/2005_warmest.html (accessed February 3, 2007); NOAA News, "1998 Warmest Year on Record, NOAA Announces," http://www.noaanews.noaa.gov/stories/s105.htm (accessed on February 3, 2007); and Science Daily, "1997 Warmest Year of Century, NOAA Reports," www.sciencedaily.com/releases/1998/01/980113062713.htm (accessed on February 4, 2007).
4. Intergovernmental Panel on Climate Change (IPCC), "Climate Change 2007: The Physical Science Basis, Summary for Policymakers, Contribution of Working Group I to the Fourth Assessment Report of the Intergovernmental Panel on Climate Change," www.ipcc.ch/SPM2feb07.pdf (accessed on February 19, 2007).
5. IPCC, *Climate Change 2001: The Scientific Basis: Contribution of Working Group 1 of the Third Assessment Report of the Intergovernmental Panel on Climate Change* (Cambridge, U.K.: Cambridge University Press, 2001).
6. National Oceanographic and Atmospheric Administration, "Global Warming: Frequently Asked Questions," available at www.ncdc.noaa.gov/oa/climate/global warming.html; National Royal Society, "Climate Science," available at www .royalsoc.ac.uk/page.asp?id=1279; and U.S. EPA, "Climate Change and Municipal Solid Waste: Two Environmental Issues With an Important Underlying Link," available at http://www.epa.gov/payt/tools/factfin.htm (accessed on June 10, 2007).
7. These examples are meant to be illustrative only. The atmospheric composition of Earth's celestial neighbors is much different than that of the Earth, but it is the concentration of greenhouse gases as much as anything that accounts for their vastly different temperatures.
8. Concentrations of atmospheric CO_2 have risen from 280 parts per million (ppm) in 1850 to 364 ppm in the late 1990s, the highest concentration on earth in twenty million years.

167

9. Respiration is the process by which plant or animal material that is consumed by humans and animals for food is broken down. It is the opposite of photosynthesis. In the digestive systems of animals, organic material is broken down and the carbon it contains is combined with oxygen and released in the form of carbon dioxide.

10. Hinrichs and Kleinbach, *Energy*.

11. Lester R. Brown, *Eco-Economy: Building an Economy for the Earth* (New York: W.W. Norton and Company, 2001).

12. IPCC, "Climate Change 2007."

13. Ibid.

14. The IPCC is an international body that was established by the World Meteorological Organization and the United Nations Environment Programme in 1988. "The role of the IPCC is to assess on a comprehensive, objective, open and transparent basis the scientific, technical and socio-economic information relevant to understanding the scientific basis of risk of human-induced climate change, its potential impacts and options for adaptation and mitigation." See *Principles Governing IPCC Work*, approved at the Fourteenth Session (Vienna, October 1–3, 1998) on October 1, 1998 and amended at the 21st Session (Vienna, November 3 and 6–7, 2003).

15. IPCC, "Climate Change 2007."

16. U.S. Environmental Protection Agency, *Inventory of U.S. Greenhouse Gas Emissions and Sinks: 1990–2000*, EPA 430-R-02-003, 2002; IPCC, *Climate Change 2001: Synthesis report, Contribution of Working Groups I, II, and III to the Third Assessment Report of the Intergovernmental Panel on Climate Change* (Cambridge, U.K.: Cambridge University Press, 2001); and IPCC, "Climate Change 2007."

17. IPCC, *Climate Change 2001*; EPA, *Inventory of U.S. Greenhouse Gas Emissions and Sinks*.

18. The unit "tons" refers to the "short tons" that are used as the unit of measurement in the United States. They should not to be confused with metric tons, which have the unit "tonnes." 1 short ton = 0.9072 metric tones.

19. U.S. Environmental Protection Agency, *Global Climate Change, Impacts for the Southeast*, a report on the September 16, 1997, EPA Regional Conference, sponsored by the EPA Office of Policy, Planning and Evaluation, Office of Economy and Environment, 1997. Available at http://yosemite.epa.gov/oar/globalwarming .nsf/UniqueKeyLookup/SHSU5BNJK2/$File/atlanta.pdf (accessed on February 17, 2006).

20. Ibid.

21. U.S. Environmental Protection Agency, "Climate Change and Arizona," 1998. Available at yosemite.epa.gov/oar/globalwarming.nsf/UniqueKeyLookup/ SHSU5BNJMV/$File/az_impct.pdf (accessed on February 17, 2007).

22. Ibid.

23. Due to the prevalence of dark building materials, restricted air flow, and lack of vegetation, urban areas tend to be warmer then the surrounding countryside and take longer to cool down during the evenings and nights. U.S. Global Change Research Program, "Preparing for a Changing Climate: The Potential Consequences of Climate Variability and Change: Southwest, 2000. Available at www.ispe.arizona.edu/research/swassess/pdf/complete.pdf (accessed

on February 17, 2007); "Impacts of Climate Change in the United States: California," www.climatehotmap.org/impacts/california.html (accessed on November 20, 2006).

24. "Impacts: California"; and U.S. Environmental Protection Agency, "Climate Change, What Does It Mean for the Southwest," 1998. Available at yosemite .epa.gov/oar/globalwarming.nsf/UniqueKeyLookup/SHSU5BPQFF/$File/dallas .pdf (accessed on February 17, 2007).

25. U.S. Environmental Protection Agency, "Climate Change and California," EPA 230-F-97-008e, 1997.

26. The California Climate Change Center, "Our Changing Climate—Assessing the Risks to California," CEC-500-2006-077, 2006. Available at www.energy.ca.gov/ 2006publications/CEC-500-2006-077/CEC-500-2006-077.PDF (accessed on February 17, 2007).

27. U.S. Global Change Research Program, "Southwest"; U.S. Environmental Protection Agency, "Climate Change and Arizona"; and "Impacts: California."

28. U.S. Environmental Protection Agency, Climate Change and Oregon, EPA 236-F-98-007u, 1998. Available at yosemite.epa.gov/oar/globalwarming.nsf/ UniqueKeyLookup/SHSU5BVJZ4/$File/or_impct.pdf (Accessed on February 17, 2007); P. W. Mote, M. Holmberg, N. J. Mantua, and Climate Impacts Group, "Impacts of Climate Variability and Change on the Pacific Northwest," 1999. Available at http://www.usgcrp.gov/usgcrp/Library/nationalassessment/pnw.pdf (accessed on February 17, 2007); and U.S. Environmental Protection Agency, "Climate Change and Washington," EPA 230-F-97-008uu, 1997. Available at yosemite.epa.gov/oar/globalwarming.nsf/UniqueKeyLookup/SHSU5BWJBX/$ File/wa_impct.pdf (accessed on February 17, 2007)

29. U.S. Environmental Protection Agency, "How Will Climate Change Affect the Mid-Atlantic Region?" EPA/903/F-00/002, 2001. Available at www.epa.gov/ maia/pdf/ClimateChange.pdf (accessed February 17, 2007); Ann Fisher, "Preparing for a Changing Climate: The Potential Consequences of Climate Variability and Change: Mid-Atlantic Overview," 2000, www.cira.psu.edu/mara/ results/foundations_report/mara.pdf (accessed March 31, 2007); and "Impacts of Climate Change in the United States: Mid-Atlantic," www.climatehotmap .org/impacts/midatlantic.html (accessed on November 30, 2006).

30. Fisher, "Preparing for a Changing Climate."

31. Great Lakes Regional Assessment Group, "Preparing for a Changing Climate: The Potential Consequences of Climate Variability and Change in the Great Lakes Region"; and "Impacts of Climate Change in the United States: Great Lakes," 2000, www.climatehotmap.org/impacts/greatlakes.html (accessed on February 3, 2007).

32. "Impacts of Climate Change in the United States: Metro East Coast," www .climatehotmap.org/impacts/metroeastcoast.html (accessed on February 4, 2007).

33. Brown, *Earth Policy Reader*.

Chapter Two: History of U.S. Climate Change Policy (pages 14–29)

1. John C. Rayson, Mayor of Pompano Beach, Florida, quoted in Seattle, www .seattle.gov/mayor/climate/quotes.htm (accessed on December 19, 2006).

2. Michael E. Kraft, *Environmental Policy and Politics* (Boston: Little, Brown, 2001).

3. Atle Christer Christiansen, "Convergence or Divergence? Status and Prospects of U.S. Climate Strategy," *Climate Policy* 3 (2003).

4. Mark Hertsgaard, *Earth Odyssey* (New York: Broadway Books, 1998).

5. Intergovernmental Panel on Climate Change (IPCC), www.ipcc.ch/about/about .htm (accessed on May 1, 2006).

6. IPCC, "Sixteen Years of Scientific Assessment in Support of the Climate Convention," 2004. Available at www.ipcc.ch/about/anniversarybrochure.pdf.

7. Ibid.

8. James Hansen, Testimony before Senate Energy and Natural Resources Committee, June 23, 1988.

9. S. Agrawala and S. Andreson, "U.S. Climate Policy: Evolution and Future Prospects, *Energy and Environment*. (cited in Armin Rosenkranz, "U.S. Climate Change Policy" *Climate Change Policy*, edited by Stephen H. Schneider, Armin Rosencranz, ed and John O. Niles, ch. 8, 221–34 (Washington, D.C.: Island Press, 2002): 222.

10. Ibid.

11. United Nations, The United Nations Framework Convention on Climate Change, Article 2. FCCC/INFORMAL/84 GE.05-62220 (E) 200705, 1992.

12. United Nations Framework Convention on Climate Change, unfccc.int/essential _background/convention/items/2627.php (accessed March 1, 2006).

13. Congressional Record, S17, 156 Daily Edition, October 7, 1992.

14. As of 2007, only Andorra, Brunei, the Holy See, Iraq, and Somalia have not signed on.

15. Armin Rosencranz, "U.S. Climate Change Policy," in *Climate Change Policy*, ed. Stephen H. Schneider, Armin Rosencranz, and John O. Niles, ch. 8, 221–34 (Washington, D.C.: Island Press, 2002).

16. Ibid.

17. Agrawala and Andresen, "U.S. Climate Policy"; and United States Senate Republican Policy Committee, rpc.senate.gov/~rpc/releases/1997/BTUcarbo-jt .htm (accessed February 17, 2007).

18. U.S. Environmental Protection Agency, "United States: Taking Action on Climate Change," 1999. Available at yosemite.epa.gov/oar/globalwarming.nsf/ UniqueKeyLookup/SHSU5BWHST/$File/usaction_99.pdf (accessed on February 17, 2007).

19. Clinton-Gore Climate Change Action Plan. Available at www.gcrio.org/ USCCAP/toc.html (accessed on February 17, 2007).

20. Rosencranz, "U.S. Climate Change Policy."

21. Leonie Haimson, "Climate Negotiation History," in *Climate Change Policy*, Appendix A, 523–29.

22. Undersecretary of State Timothy Wirth at COP-2 in Geneva in July 1996, cited in M. Grubb, C. Vrolijk, and D. Brack, *The Kyoto Protocol: A Guide and Assessment* (London: Royal Institute of International Affairs, 2000), 54.

23. Agrawala and Andresen, "U.S. Climate Policy."

24. Congressional Record, July 27, 1997, S8113–8138.

25. For a complete list of reduction targets, see Appendix A.

26. Haimson, "Climate Negotian History."

27. For a complete list of targets, see Appendix A.

28. Joint implementation programs and the clean development mechanism are both market-based strategies to reduce emissions at minimal cost. They are described in more detail later in this chapter, in the section on federal policy options.

29. Haimson, "Climate Negotiation History."

30. In the United States, the President has the power to sign international treaties, but they do not go into effect as binding until they are ratified by the Senate. Therefore, without Senate approval, President Clinton's act in signing the Kyoto Protocol was only symbolic.

31. Agrawala and Andresen, "U.S. Climate Policy."

32. "Hotting Up In the Hague," *The Economist,* November 18, 2000, 83.

33. President Bill Clinton, State of the Union Address, January 27, 2000.

34. The White House, Office of the Press Secretary, Press Release, February 3, 2000. Available at http://clinton4.nara.gov/WH/New/html/20000204_8.html (accessed on February 3, 2007).

35. Letter to *New York Times* dated January 8, 2001, from White House Chief of Staff John Podesta. *New York Times,* January 10, 2001, A22.

36. "Letter from the President to Senators Hagel, Helms, Craig, and Roberts," The White House, Office of the Press Secretary, March 13, 2001.

37. Press briefing at the White House, March 29, 2001.

38. "Fact Sheet: President Bush Announces Clear Skies and Global Climate Change Initiatives," White House Office of the Press Secretary, www.whitehouse.gov/news/releases/2002/02/print/20020214.html (accessed February 17, 2007); and "Bush to Unveil Plan Linking Economy and Environment," *Wall Street Journal,* February 14, 2002, A22, cited in Rosencranz, "U.S. Climate Change Policy."

39. "President Bush Announces Clear Skies."

40. A. C. Revkin and K. Q. Seelye, "Report by EPA Leaves Out Data on Climate Change," *New York Times,* June 19, 2003.

41. Strategic Plan of the U.S. Climate Change Science Program, www.climatescience.gov/Library/stratplan2003/default.htm (accessed February 17, 2007).

42. Pew Center on Global Climate Change, "Summary of The Lieberman-McCain Climate Stewardship Act," www.pewclimate.org/policy_center/analyses/s_139_summary.cfm (accessed on February 17, 2007).

43. The Protocol passed its first major hurdle of meeting the requirement that it be ratified by fifty-five countries when Iceland ratified the treaty on May 23, 2002, but it still could not go into effect until the countries that ratified the Protocol accounted for 55% of the world's greenhouse gas emissions.

44. The joint meeting of the 11th Conference of Parties (COP) (i.e., signatory nations to the Kyoto Protocol) and the 1st Meeting of the Parties (MOP).

45. Amanda Griscom Little, "On the Right Track: New Republican Leaders Emerging in Battle against Climate Change," *Grist Magazine,* February 4, 2005.

46. Congressional Record—Senate, June 22, 2005, S7033–S7037; and American Institute of Physics, "Shift in Senate Thinking on Climate Change," *FYI: The AIP Bulletin of Science Policy News* 114 (July 26, 2005), www.aip.org/fyi/2005/114.html (accessed on February 18, 2007).

47. American Geologic Institute, Government Affairs Program, "Summary of Hearings on Climate Change," www.agiweb.org/gap/legis109/climate_hearings.html (accessed on December 7, 2006).

48. Senate Floor Statement by U.S. Senator James M. Inhofe (R-OK), January 4, 2005, inhofe.senate.gov/pressreleases/climateupdate.htm (accessed on December 7, 2006).
49. Supreme Court of the United States, *Massachusetts et al. v. Environmental Protection Agency et al.* No. 05–1120. Argued November 29, 2006—Decided April 2, 2007, www.supremecourtus.gov/opinions/06pdf/05-1120.pdf (accessed June 10, 2007).
50. Julian Borger, David Adam, Suzanne Goldenberg, "Bush Kills Off Hopes For G8 Climate Change Plan," *The Guardian Unlimited,* June 1, 2007, environment.guardian.co.uk/climatechange/story/0,,2093055,00.html.
51. N. J. Vig and M. E. Kraft, eds., *Environmental Policy: New Directions for the Twenty-first Century,* 5th ed. (Washington, DC: CQ Press, 2003).

Chapter Three: Local Actions, Global Results (pages 30–45)

1. Kenneth E. Patton, Mayor of Brooklyn, Ohio, quoted in Seattle, www.seattle .gov/mayor/climate/quotes.htm (accessed on December 19, 2006).
2. See Chapter One, "Effects of a Changing Climage in the United States" for a discussion of the potential impacts around the country.
3. Institute for the Study of Society and Environment, www.isse.ucar.edu/ newshp95_4/pastmeet.html (accessed on December 5, 2006).
4. The complete text of this declaration has been included in Appendix B.
5. Numbers reported by the City of Seattle on January 10, 2007. Current information available at www.seattle.gov/mayor/climate/quotes.htm (accessed on January 7, 2007).
6. Text from these commitments taken from City of Seattle's website on the U.S. Mayor's Climate Protection Agreement, www.seattle.gov/mayor/climate/ (accessed on January 20, 2007). More information on this resolution and its signatories is available in Appendix C.
7. This has since become an annual event. The second Sundance Summit was held in November 2006. More information on these events can be found at www .sundancesummit.org.
8. For more sample events and programs see www.iclei.org/index.php?id=448 (accessed on February 17, 2007).
9. Formerly known as the International Council for Local Environmental Initiatives.
10. ICLEI Global Programs, Cities for Climate Protection, www.iclei.org/index.php? id=800 (accessed on January 14, 2007).
11. Personal conversation with Susan Ode, ICLEI Membership Officer, December 2006. For a complete list of communities involved in the CCP, see Appendix D.
12. For more information, see the ICLEI U.S. website, www.iclei.org/us.
13. ICLEI—Local Governments for Sustainability, "ICLEI U.S. Cities for Climate Protection Progress Report," 2005. A discussion of ICLEI's quantification methodology can be found in Appendix E.
14. Pew Center on Global Climate Change, "U.S. Greenhouse Gas Emissions, 1990–2004, www.pewclimate.org/docUploads/USEmissions2004%5FFeb06 %2Epdf (accessed on February 17, 2007).
15. American Lung Association (ALA), "Nitrogen Dioxide," 2000, www.lungusa .org/site/pp.asp?c=dvLUK9O0E&b=35355 (accessed on February 17, 2007).

16. ICLEI, "ICLEI U.S. Cities,"
17. ALA, "Nitrogen Dioxide."
18. Ibid.
19. United States Environmental Protection Agency (EPA), "An Introduction to Indoor Air Quality—Nitrogen Dioxide," www.epa.gov/iaq/no2.html (accessed on February 17, 2007).
20. American Lung Association (ALA), "Sulfur Dioxide," 2000, www.lungusa.org/site/pp.asp?c=dvLUK9O0E&b=35358 (accessed on February 17, 2007).
21. Ibid.
22. United States Environmental Protection Agency (EPA), "Health and Environmental Impacts of SO_2," www.epa.gov/air/urbanair/so2/hlth1.html (accessed on February 17, 2007).
23. Ibid.
24. American Lung Association (ALA), "Carbon Monoxide," 2000, www.lungusa.org/site/pp.asp?c=dvLUK9O0E&b=35332 (accessed on February 17, 2007).
25. Ibid.
26. Ibid.
27. United States Environmental Protection Agency (EPA), "An Introduction to Indoor Air Quality—Carbon Monoxide," www.epa.gov/iaq/co.html (accessed on February 17, 2007).
28. United States Environmental Protection Agency (EPA), "An Introduction to Indoor Air Quality—Organic Gases," www.epa.gov/iaq/voc.html (accessed on February 17, 2007).
29. Ibid.
30. For comparison, a human hair is about 75 microns in diameter.
31. American Lung Association (ALA), "Particulate Matter," 2000, www.lungusa.org/site/pp.asp?c=dvLUK9O0E&b=35356 (accessed on February 17, 2007).
32. Ibid.
33. ICLEI, "ICLEI U.S. Cities."
34. EnergyStar, "2004 Annual Report," www.energystar.gov/ia/news/downloads/annual_report2004.pdf (accessed on February 17, 2007).
35. Ibid.
36. B. Hopkins, "Renewable Energy and State Economies. Trend Alert: Critical Information for State Decision-Makers," 2003, www.csg.org/pubs/Documents/TA0305RenEnergy.pdf (accessed on February 17, 2007).
37. Texas Transportation Institute, "2005 Urban Mobility Report," mobility.tamu.edu/ums/report (accessed on December 27, 2005).

Chapter Four: Constraints on Local Policymakers (pages 46–50)

1. Mike McKinnon, Mayor of Lynnwood, Washington, quoted in Seattle, www.seattle.gov/mayor/climate/quotes.htm (accessed on December 19, 2006).
2. Michele M. Betsill, "Localizing Global Climate Change: Controlling Greenhouse Gas Emissions in U.S. Cities," Belfer Center for Science and International Affairs (BCSIA) Discussion Paper 2000-20, Environmental and Natural Resources Program, Kennedy School of Government, Harvard University.
3. Ibid.

4. Ibid.
5. John W. Kingdon, *Agendas, Alternatives and Public Policies* (Boston: Little, Brown, 1984).
6. John Immerwahr, "Waiting for a Signal: Public Attitudes toward Global Warming, the Environment and Geophysical Research," *Public Agenda*, April 15, 1999.
7. Ibid.
8. Fox News, www.foxnews.com/story/0,2933,175070,00.html (accessed on February 17, 2007).
9. Immerwahr, "Waiting for a Signal"; and ibid.
10. Naomi Oreskes, "Beyond the Ivory Tower: The Scientific Consensus on Climate Change," *Science* (December 2004).
11. Maxwell T. Boykoff and Jules M. Boykoff, "Balance as Bias: Global Warming and the U.S. Prestige Press," *Global Environmental Change Part A* 14, no. 2 (July 2004).
12. Betsill, "Localizing Global Climate Change."
13. Ibid.
14. Ibid.
15. Ibid.; and Daniel Mazmanian and Paul A. Sabatier, "The Implementation of Public Policy: A Framework for Analysis," *Policy Studies Journal* 8 (1980): 538–60.
16. Daniel Press, "Local Environmental Policy Capacity: A Framework for Research," *National Resources Journal* 38 (Winter 1998): 29–52.
17. Betsill, "Localizing Global Climate Change."
18. Mazmanian and Sabatier, "Implementation of Public Policy."

Chapter Five: Global Issues, Local Action (pages 51–68)

1. Dave Cieslewicz, Mayor of Madison, Wisconson, quoted in Seattle, www.seattle.gov/mayor/climate/quotes.htm (accessed on December 19, 2006).
2. Carolyn Kousky and Stephen H. Schneider, "Global Climate Policy: Will Cities Lead the Way? *Climate Policy* 3 (2003).
3. Harriet Bulkeley and Michelle M. Betsill, *Cities and Climate Change* (London: Routledge, 2003).
4. Michael E. Kraft, *Environmental Policy and Politics* (Boston: Little, Brown, 2001).
5. Kousky and Schneider, "Global Climate Policy."
6. United States Green Building Council, LEED-2.1 Reference Guide.
7. William J. Fisk, "Review of Health and Productivity Gains From Better IEQ," LBNL-48218, Environmental Energy Technologies Division, Indoor Environment Department, Lawrence Berkeley National Laboratory, Berkeley, California, (available at repositories.calif.org/lbnl/LBNL-48218).
8. For more information, see the City of Berkeley's website, "Best Practices for Climate Protection: A Local Government Guide," www.ci.berkeley.ca.us/sustainable/residents/RECO/RECO_Guide2Compliance.html and International Council for Local Environmental Initiatives.
9. For more information see www.ci.berkeley.ca.us/sustainable/buildings/ceco.html

10. University of California—Santa Barbara, "Sustainable Environmental Indicators," www.es.ucsb.edu/proj/135Bindicators2/santabarbara/environmental/green buildings.html (accessed on February 3, 2007).

11. International Council for Local Environmental Initiatives (ICLEI), *Best Practices for Climate Protection: A Local Government Guide* (Oakland, Calif.: ICLEI, 2000).

12. Mary Alice Kaspar, "Buried Treasure: Austin to Gain New Energy Source from Landfill," *Austin Business Journal*, June 11, 2004.

13. Ibid.

14. Oregon State Department of Energy, Renewable Energy Division, www.oregon.gov/ENERGY/RENEW/Biomass/FuelCell.shtml (accessed on February 11, 2007).

15. ICLEI, *Best Practices*.

16. Austin Energy, "Green Choice® Austin Energy's Renewable Energy Program," www.austinenergy.com/Energy%20Efficiency/Programs/Green%20Choice/index.htm (accessed on February 11, 2007).

17. United States Environmental Protection Agency, National Center for Environmental Assessment, "Draft Report on the Environment 2003," www.epa.gov/indicators (accessed on February 17, 2007).

18. United States Environmental Protection Agency, Office of Solid Waste, "Municipal Solid Waste in the United States: 2005 Facts and Figures," www.epa.gov/msw/pubs/ex-sum05.pdf (accessed on February 17, 2007).

19. An earlier version of this recycling example originally appeared in Ryan Bell and Rozy Fredericks, *Waste Prevention and Sustainability: Case Studies for Local Governments* (ICLEI—Local Governments for Sustainability, 2005).

20. City of Seattle, Seattle Public Utilities, www.seattle.gov/util/About_SPU/Recycling_System/History_&_Overview/Ban_on_Recyclables_in_Garbage/index.asp (accessed on February 9, 2007).

21. Yard debris has been prohibited since 1989. The ordinance also contains an exception for paper and cardboard that is not recyclable due to contamination by food or other materials.

22. City of Seattle. Administrative Rule #SPU-DR-01-04: Prohibition of Recyclables in Garbage.

23. "Waste Prevention and Sustainability: Case Studies for Local Governments," compiled for StopWaste.Org by ICLEI—Local Governments for Sustainability.

24. For instance, in the early stages of the policy's development, hotels raised concerns about the ban and their ability to ensure guest participation. The city worked with them to develop a plan for them to reach their clients in order to ensure their compliance.

25. Compliance estimated based on the number of waste containers that have not been collected; waste haulers are not required to open sealed bags to check for recyclables, therefore this could be an overestimation. Financial savings reported in Sharon Pian Chan, "Seattle Getting Better at Recycling its Trash," *Seattle Times*, November 25, 2006.

26. United States Environmental Protection Agency, Office of Solid Waste, Municipal Solid Waste in the United States: 2005 Facts and Figures," www.epa.gov/msw/pubs/ex-sum05.pdf (accessed on February 17, 2007).

27. State of Oregon, Land Quality Division, Commercial Waste Reduction Clearinghouse, www.deq.state.or.us/lq/sw/cwrc/index.htm (accessed on February 10, 2007).

28. United States Environmental Protection Agency, Solid Waste and Emergency Response Division, "Cutting the Waste Stream in Half Community Record-Setters Show How," EPA-530-F-99-017, 1999; and ICLEI, *Best Practices*.

29. Twin Cities Free Market, www.twincitiesfreemarket.org (accessed on February 9, 2007).

30. Excess Access, "Excess Access Statistics," www.excessaccess.com/pages/main_pressstats806.html (accessed on February 19, 2007).

31. American Council for an Energy Efficient Economy, www.aceee.org/press/ie054pr.htm (accessed on September 11, 2005).

32. Ibid.

33. Assuming a 3.4-gallon savings per flush and an average ten flushes per day per household.

34. United States Department of Energy, Energy Information Administration, "Retail Gasoline Historical Prices," www.eia.doe.gov/oil_gas/petroleum/data_publications/wrgp/mogas_history.html (accessed on January 6,2007). Based on the average price of all grades, all formulations of gasoline.

35. This assumes a switch from a midsized vehicle that gets 22 mpg to a hybrid that gets 56 mpg (combined city and highway fuel efficiency) and that 12,000 miles are driven in a typical year.

36. Bureau of Transportation, www.bts.gov/publications/national_transportation_statistics/html/table_01_11.html (accessed February 19, 2007).

37. City of Santa Monica, Transportation Management, www.santa-monica.org/planning/transportation/abouttransmanagementtmo.html (accessed on February 6, 2007); and ICLEI, *Best Practices*.

38. City of Seattle, Bicycle Program, www.seattle.gov/transportation/bikeprogram.htm (accessed on February 11, 2007).

39. International Council for Local Environmental Initiatives. Best Practices for Climate Protection: A Local Government Guide

40. U.S. Climate Change Technology Program, "U.S. Climate Change Technology Program: Technology Options for the Near and Long Term," www.climatetechnology.gov/library/2005/tech-options/index.htm (accessed February 18, 2007).

41. As calculated by Ryan Bell using the ICLEI CCP methodology based on data provided by the City of Tucson.

42. Environmental Protection Agency, "Chicago's Heat Island Reduction Activities," www.epa.gov/heatislands/pilot/chic_activities.html (accessed on February 11, 2007).

43. Emissions savings calculated by Ryan Bell using the ICLEI CCP methodology. Data for the section and subsequent calculations were gathered from personal conversations and from City of Chicago, Department of the Environment, "Monitoring the Rooftop Garden's Benefits," and "About the Rooftop Garden," egov.cityofchicago.org/city/webportal/portalDeptCategoryAction.do?deptCategoryOID=-536889314&contentType=COC_EDITORIAL&topChannelName=Dept&entityName=Environment&deptMainCategoryOID=-536887205 (accessed on February 18, 2007).

44. Lawrence Berkeley National Laboratory, Environmental Energy Technologies Division, Heat Island Group, "Cool Pavements," eetd.lbl.gov/HeatIsland/Pavements/ (accessed February 18, 2007).

45. James Simpson, "Urban Forest Impacts on Regional Cooling and Heating Energy Use: Sacramento County Case Study," *Journal of Arboriculture* 24, no. 4 (1998): 201–14.

46. Gregory McPherson and James Simpson, *Carbon Dioxide Reduction through Urban Forestry*, Gen Tech Rep PSW-GTR-171 (Albany, Calif.: Pacific Southwest Research Station, USDA, US Forest Service, 1999).

47. Miami-Dade County Cool Communities Program.

48. K.I. Scott, James Simpson, and Gregory McPherson, "Effects of Tree Cover on Parking Lot Microclimate and Vehicle Emissions," *Arboric* 25, no. 3 (1999).

49. City of Sacramento, Parking Lot Tree Shading Design and Maintenance Guidelines, 2002.

Chapter Six: Governments Gone Green (pages 69–90)

1. John Medinger, Mayor of La Crosse Wisconson, quoted in Seattle, www.seattle.gov/mayor/climate/quotes.htm (accessed December 19, 2006).

2. United States Green Building Council, www.usgbc.org/DisplayPage.aspx?CategoryID=1 (accessed on February 19, 2007).

3. See Appendix F for a complete list of state and local governments that have adopted an LEED ordinance.

4. USGBC, "Introduction to LEED and Green Building powerpoint," www.usgbc.org/ShowFile.aspx?DocumentID=742#300,9,Slide 9 (accessed on February 19, 2007).

5. Ibid.

6. Tucson example from International Council for Local Environmental Initiatives (ICLEI), *Best Practices for Climate Protection: A Local Government Guide*.

7. Al Nichols Engineering Inc., "Energy Use in Civano and Tucson Residences," prepared for the community of Civano LLC, 2003, www.civanoneighbors.com/docs/environment/Civano_EnergyFnl03.pdf (accessed on February 5, 2007).

8. Toledo example from ICLEI, *Best Practices*.

9. City of Santa Monica, "Ten Ways Santa Monica is Working to Address Climate Change" www.smgov.net/epd/news/pdf/GWarming_Response.pdf (accessed on February 11, 2007).

10. United States Department of Energy, "Energy Efficiency and Renewable Energy," www.eere.energy.gov/greenpower/buying/customers.shtml?page=1&companyid=27 (accessed on February 17, 2007); and ICLEI, *Best Practices*.

11. City of Santa Monica, www.smgov.net/news/releases/archive/2004/epwm20040218.htm (accessed on February 11, 2007).

12. United States Department of Energy, "Energy Efficiency and Renewable Energy," www.eere.energy.gov/greenpower/buying/customers.shtml?page=1&companyid=27 (accessed on February 11, 2007).

13. ICLEI, *Best Practices*.

14. United States Department of Energy, "Energy Efficiency and Renewable Energy," www.eere.energy.gov/greenpower/buying/customers.shtml?page=1&companyid=

327 (accessed on February 11, 2007). Calculation performed using the ICLEI's Clean Air Climate Protection Software.

15. An earlier version of this biodiesel example originally appeared in Ryan Bell and Rozy Frederick, *Waste Prevention and Sustainability: Case Studies for Local Governments* (ICLEI—Local Governments for Sustainability, 2005).

16. To see the network of fueling stations, go to www.biodiesel.org/buyingbiodiesel/ retailfuelingsites/default.shtm.

17. United States Environmental Protection Agency, Biodiesel Emissions Analysis Program, with link to "Comprehensive Analysis of Biodiesel Impacts on Exhaust Emissions," October 2002, www.epa.gov/otaq/models/biodsl.htm (accessed on February 17, 2007).

18. United States Environmental Protection Agency, "Comprehensive Analysis of Biodiesel Impacts on Exhaust Emissions," draft technical report, October 2002, www.epa.gov/otaq/models/analysis/biodsl/p02001.pdf. See also R. E. Morris et al., "Impact of Biodiesel Fuels on Air Quality and Human Health: Summary Report, September 16, 1999–January 31, 2003," National Renewable Energy Lab, Golden, Colorado, www.nrel.gov/docs/fy03osti/33793.pdf.

19. J. Sheehan et al., "Lifecycle Inventory of Biodiesel and Petroleum Diesel for Use in an Urban Bus," National Renewable Energy Laboratory for the U.S. Department of Energy Office of Fuels Development and the U.S. Department of Agriculture Office of Energy, Golden, Colorado.

20. National Biodiesel Board, "Biodiesel Emissions Fact Sheet," www.biodiesel.org/ resources/fuelfactsheets/default.shtm (accessed on February 17, 2007).

21. From the May 2001 Clean Cities Fact Sheet on biodiesel for local governments, www.eere.energy.gov/afdc/pdfs/Biodiesel-f3.pdf.

22. Data taken from the U.S. Department of Energy, Alternative Fuels Data Center, "Alternative Fuel Price Reports," 2000 through 2006.

23. A 1998 amendment to the Energy Policy Act allows regulated federal, state, and private fleet operators to meet up to 50% of their alternative-fueled vehicle purchase requirements by switching to biodiesel. For more information, see the National Biodiesel Board's fact sheet, www.biodiesel.org/pdf_files/fuelfactsheets/ EPACTSummary.PDF.

24. United States Department of Energy, Energy Efficiency and Renewable Energy, Clean Cities Coalition, "Success Story: Las Vegas Serves up Biodiesel for Local Schools," 2004, www.eere.energy.gov/afdc/apps/toolkit/pdfs/las_vegas _success.pdf.

25. Although some virgin oil needs to be imported to meet the region's growing demand.

26. Without this additive, it is estimated that this switch to biodiesel would increase NO_x emissions by 5 tons annually.

27. City of Denver, "Sustainable Initiatives," www.denvergov.org/ES/SustainableIni tiatives/tabid/386886/Default.aspx (accessed on February 12, 2007); and ICLEI, *Best Practices*.

28. City of Denver, "Sustainable Initiatives."

29. Ibid.

30. Greenprint Denver, www.greenprintdenver.org/energy/greenfleet.php (accessed on February 12, 2007).

31. ICLEI, *Best Practices*.

32. Sierra Club, "Cool Cities: Solving Global Warming One City at a Time," 2005, www.coolcities.us/files/coolcitiesguide.pdf (accessed on February 15, 2007).
33. $400/$3/gallon = 133.3 gallons × 22 mpg × 20 lbs CO_2/2,000 lbs per ton.
34. ICLEI, *Best Practices*.
35. An earlier version of this Portland example originally appeared in Bell and Fredericks, *Waste Prevention and Sustainability*.
36. City of Portland, "Sustainable Paper Use Policy," www.portlandonline.com/shared/cfm/image.cfm?Id = 24521 (accessed on February 5, 2007).
37. Calculation based on a 100% methane recovery rate, using the mixed paper emissions reduction factors, and assuming that all reduced paper consumption otherwise would have been landfilled.
38. An earlier version of these Miami-Dade County examples originally appeared in Bell and Fredericks, *Waste Prevention and Sustainability*.
39. Miami-Dade County Department of Environmental Resources, "A Long-Term Co_2 Reduction Plan for Miami-Dade County, Florida, 1993–2006," December 12, 2006, www.miamidade.gov/derm/globalwarminglibrary/Co2-Reduction-Final-Report.pdf.
40. Example from ICLEI, *Best Practices*.
41. An earlier version of this "environmentally preferable purchasing" example originally appeared in Bell and Fredericks, *Waste Prevention and Sustainability*.
42. Federal Executive Order 13101, "Greening the Government through Waste Prevention, Recycling, and Federal Acquisition," www.epa.gov/oppt/epp/pubs/13101.pdf (accessed on February 17, 2007).
43. Federal Register, "Part VII Environmental Protection Agency: Final Guidance on Environmentally Preferable Purchasing for Executive Agencies," *Notice* 64, no. 161, August 20, 1999, pp. 45816–25.
44. U.S. Department of Energy, Energy Information Administration, "Emissions of Greenhouse Gases in the United States, 1997"; A. Hendriks, E. Worrell, D. de Jager, K. Blok, and P. Riemer, "Emission Reduction of Greenhouse Gases from the Cement Industry," Greenhouse Gas Control Technologies Conference paper, www.tececo.com/files/Sustainability_Documents/EmissionReductionofGreenhouse-GasesfromtheCementIndustry.pdf (accessed on June 6, 2005); P. K. Mehta, "Role of Pozzolanic and Cementitious By-Products in Sustainable Development of the Concrete Industry," in Sixth CANMET/ACI/JCI Conference: Fly Ash, Silica Fume, Slag and Natural Pozzolans in Concrete, Tokushima, Japan, 1998. An earlier version of the recycled content cement example originally appeared in Bell and Fredericks, *Waste Prevention and Sustainability*.
45. California Integrated Waste Management Board, "California Green Roads," www.zerowaste.ca.gov/RCM/pdf/Folder.pdf (accessed on November 22, 2006).
46. California Energy Commission, "California 'Green Lights' Energy Savings with New Traffic Signals," 2002, www.energy.ca.gov/releases/2002_releases/2002-03-14_led_signals.html (accessed on February 19, 2007).
47. ICLEI, *Best Practices*.
48. Ibid.
49. Ibid.
50. Ann Arbor case study excerpted from and unpublished case study produced by ICLEI—Local Governments for Sustainability.

Chapter Seven: Building a Climate Action Policy (pages 91–106)

1. Rocky Mountain Climate Organization, www.rockymountainclimate.org/release
   _launching_2.htm (accessed on January 8, 2006).
2. Natural Gas Vehicle Network, www.ngvnetwork.com/news/04/march8-14.htm
   (accessed on December 29, 2006).
3. City of Fort Collins, Resolution 97-97 of the Council of the City of Fort Collins
   Authorizing the City of Fort Collins to Join the Cities for Climate Protection
   Campaign, passed July 1, 1997, www.fcgov.com/climateprotection/ccpresolution
   .php (accessed on February 11, 2007).
4. Ibid.
5. The list has been summarized, from John F. Fischbach and Lucinda R. Smith,
   "Climate Protection: Fort Collins Likes the Idea," reprint of article in the August
   2000 issue of *Public Management,* a monthly publication of the International
   City/County Management Association, www.fcgov.com/climateprotection/pma-
   final.php (accessed on September 1, 2005).
6. Analysis and text for this section provided by ICLEI—Local Governments for
   Sustainability, originally included in their Milestone Guide.
7. Data for this section taken from the City of Fort Collins, Natural Resources Department, "City of Fort Collins Local Action Plan to Reduce Greenhouse Gas
   Emissions," 1999.
8. City of Fort Collins, "Local Action Plan."
9. Sample measures in this section are from City of Fort Collins, "Local Action
   Plan." A complete list of measures included in the action plan are included in
   Appendix G. Summarization and text by ICLEI—Local Governments for Sustainability.
10. City of Fort Collins Energy Management Team, "City of Fort Collins 2003/2004
    Climate Protection Status Report."

Chapter Eight: Achieving Results (pages 107–18)

1. Portland Office of Sustainable Development, Multnomah County Department
   of Sustainable Community Development, "Local Action Plan on Global Warming: City of Portland and Multnomah County," 2001.
2. "A Progress Report on the City of Portland and Multnomah County Local Action Plan on Global Warming: June 2005."
3. Ibid.
4. Governor Tom McCall, Legislative Address, 1973. Oregon Secretary of State,
   Oregon State Archives, arcweb.sos.state.or.us/governors/McCall/legis1973.htm
   (accessed on March 31, 2007).
5. The Urban CO_2 Reduction Initiative was the pilot project that gave rise to the
   protocols and methodologies that eventually were embodied in the Cities for
   Climate Protect (CCP) Campaign (see chapter Three).
6. City of Portland, "Global Warming Reduction Strategy," adopted November
   2003, www.portlandonline.com/shared/cfm/image.cfm?Id=112110 (accessed on
   June 12, 2007).
7. Portland Office of Sustainable Development, "Local Action Plan."

8. T. M. Houghton, C. F. Jenkins, and J. J. Ephramus, eds., *Climate Change: The IPCC Scientific Assessment* (Intergovernmental Panel on Climate Change, Cambridge University Press, 1990).

9. Forty-four million kilowatts is enough electricity to meet the energy needs of four thousand homes for a year.

10. Portland Energy Office, "City of Portland 1990 Energy Policy: Impacts and Achievements," 2000; and Portland Energy Office, "City of Portland Carbon Dioxide Reduction Strategy."

11. "A Progress Report on the City of Portland and Multnomah County Local Action Plan on Global Warming: June 2005."

12. Michael Armstrong, City of Portland, Oregon, Office of Sustainable Development, personal communication, June 6, 2007.

13. City of Portland, Office of Sustainability, Energy Division, "Portland Climate Change Efforts," April 2003, www.portlandonline.com/shared/cfm/image.cfm? Id=112117 (accessed on February 17, 2007).

14. CNN Money, "The Best Places to Live," money.cnn.com/magazines/moneymag/moneymag_archive/2000/12/01/292674/index.htm (accessed on February 17, 2007).

15. Ibid.

16. The bullet points were excerpted from "A Progress Report on the City of Portland and Multnomah County Local Action Plan on Global Warming: June 2005."

17. City of Portland, "Portland Receives 2002 Climate Protection Award," www.portlandonline.com/mayor/index.cfm?&a=bfffa&c=cifgb (accessed on November 27, 2004).

18. "A Progress Report on the City of Portland and Multnomah County Local Action Plan on Global Warming: June 2005."

19. Director of Strategic Planning Abby Young, ICLEI—Local Governments for Sustainability Personal Conversation, 2005.

Chapter Nine: Warming to the Future (pages 119–23)

1. Greg Nickels, Mayor of Seattle, Washington, quoted in Seattle, www.seattle.gov/mayor/climate/quotes.htm#seattle (accessed on December 19, 2006).

2. According to the IPCCs website, over 2,500 scientific expert reviewers, 850 contributing authors, and 450 lead authors, representing 130 countries, contributed to this report and its findings. See www.ipcc.ch/ (accessed on February 19, 2007).

3. "Very likely" indicates that the authors are 90% confident that this is indeed the case.

4. Intergovernmental Panel on Climate Change (IPCC), "Climate Change 2007: The Physical Science Basis," Summary for Policymakers, Contribution of Working Group I to the Fourth Assessment Report of the Intergovernmental Panel on Climate Change, 2007, www.ipcc.ch/SPM2feb07.pdf (accessed on February 19, 2007).

About the Authors

Tommy Linstroth is the head of Sustainable Initiatives for Melaver, Inc., a vertically integrated, sustainable real estate firm based in Savannah, Georgia. Tommy is responsible for spearheading all of the company's sustainable real estate developments, including one of the first LEED-certified projects in the United States to be cited on the National Historic Register and the first all-retail LEED-certified development in the country. Since then, Tommy has taken responsibility for adding another thirteen sustainable projects throughout the Southeast to the company's pipeline, with a market value of roughly a quarter of a billion dollars—making him responsible for one of the largest sustainable portfolios in the United States.

In addition to making sure that the company develops in ways that meet the triple bottom line (returns that are profitable not only financially, but also socially and ecologically), Tommy works to reduce the impacts of Melaver, Inc.'s business operations—efforts that have led to the company achieving climate neutrality and inclusion in the EPA's Climate Leaders' Program and Green Power Partnership. He is a regular speaker on sustainable development in Georgia and throughout the country.

Tommy is the founder and president of the Savannah Chapter of the U.S. Green Building Council, a 501c3 organization dedicated to improving the built environment in Savannah and the coastal Southeast, a member of the board of trustees of the Live Oak Public Libraries and of the board of directors for the Georgia Conservation Voters Education Fund.

A native of Racine, Wisconsin, Tommy received his B.S. in Business Administration from the University of Wisconsin—Platteville in 2000 and his M.S. in Environmental Studies from the College of Charleston in 2004, where he focused on the ability of municipalities to address climate change.

• • •

Ryan Bell is currently an Environmental Planner with the Bay Area Air Quality Management District, where he works with organizations, communities, and individuals in the greater San Francisco Bay region to reduce emissions of common air pollutants, particularly from vehicles and other mobile sources. He also has served on various advisory committees for municipal and nonprofit organizations.

Prior to his position with the Air District, Ryan served as the U.S. Program Manager for ICLEI's Cities for Climate Protection campaign and the Communities 21 sustainability campaign. With the CCP campaign, Ryan managed ICLEI's urban heat-island reduction initiative, oversaw the formation of regional emissions reduction networks, organized local and international emission reduction and sustainability conferences and workshops, and led ICLEI's quantification work and technical assistance program. As coordinator of ICLEI's sustainability initiatives, Ryan oversaw the development of sustainability inventories, authored best practices guides, and was a regular speaker at sustainability conferences, discussing sustainability indicators and the actions that local governments can take to improve environmental, economic, and social conditions in their communities.

Previously, Ryan worked for the City of Olympia as coordinator of the city's Climate and Sustainability Workgroup, assistant to the Commute Trip Reduction program, and researcher on land-use development patterns to maximize the air and water impacts provided by urban forests. He also worked for three years with communities in Eastern Europe, as both a Peace Corps volunteer and as an independent consultant to preserve their plant and animal species and reduce their emission levels.

Ryan received his M.S. in environmental studies from The Evergreen State College, where he emphasized environmental policy and community development and he holds a Bachelor of Science degree in conservation biology from Brigham Young University.

Index

Acid rain, 38, 39

Agriculture, 7, 17

Air quality: biodiesel's effect on, 76, 77; and emissions reduction, 37–41, 64, 94; federal support for, 20, 89; and green municipal fleet, 79; and public health, 37–41; and renewable energy sources, 74

Alternative energy sources: biofuels, 76–78; and green buildings, 70; methane from decomposition, 55–56, 113; for municipal operations, 74–78; Seattle's commitment to, 34. *See also* Renewable energy sources

Aluminum industry, 17, 42–43; and bauxite mining and manufacture, 42

Ann Arbor, Michigan, public transportation, 89

Anthropogenic sources of greenhouse gasses, 1–2, 4–8, 15, 119–20

Arrhenius, Svante, 5–6

Atmospheric gases, climatic role of, 2–5, 167nn7–8

Austin Energy, 55

Australia, 22

Automobiles. *See* Vehicles

Bergen County, New Jersey, waste management, 59

Berkeley, California, emissions reduction, 52–53, 86

Berlin Mandate, 17

Bicycling promotion, 65–66, 113

Binding commitments to emissions reduction, 17–19

Bingaman, Senator Jeff, 24

Biofuels, 33, 56, 76–78, 112–13, 178n23

Blended cement, 86–87

Boxer, Senator Barbara, 24, 25

Building codes as incentive tools, 52

Buildings and construction: acid rain's threat to, 39; cool materials, 66–67; EnergyStar Buildings partnership program, 41–42; LEED initiatives list, 157–58; private sector green initiatives, 33, 52–54, 111, 112, 116; public sector green initiatives, 69–74, 111, 112, 116

Bureaucracy challenge to emissions reduction, 48–49

Bush, Pres. George H. W., 14–15, 15

Bush, Pres. George W., 20; administration on climate change, 20–23

Businesses: farming industry's opposition to emission control, 17; green buildings for, 53–54; municipal partnership with, 103–6; recycling programs, 58–59, 113; Seattle's commitment to more efficient, 33

Byrd-Hagel resolution, 18, 19

California: Berkeley's programs, 52–53, 86; environmental leadership, 26, 67; Los Angeles's programs, 63, 74; San Diego's programs, 63; Santa Barbara's programs, 54; Santa Monica's programs, 64, 74

Cancer-causing emissions, 40

Cap-and-trade programs: in California, 26; definition, 22; federal legislation for, 24; in Fort Collins, 94; historical development, 18; industrialized nations' desire for, 19; in Kyoto Protocol, 28; and level playing field issue, 122; state and regional level, 26, 123

Carbon cycle, 3–5

Carbon dioxide (CO_2): anthropogenic emissions, 6–7; atmospheric role of, 3–4; cement manufacturing release of, 86; and GWP, 8, 9; rise in atmospheric concentration, 5, 167n8